101 PATENTED SOLAR ENERGY USES

101 PATENTED SOLAR ENERGY USES

DANIEL J. O'CONNOR

VAN NOSTRAND REINHOLD COMPANY
New York Cincinnati Toronto London Melbourne

Library of Congress Catalog Card Number 80-13773
ISBN 0-442-24432-0 (pbk.)

Printed in the United States of America.

Designed by Jean Callan King/Visuality

Published by Van Nostrand Reinhold Company
135 West 50th Street, New York, NY 10020

Van Nostrand Reinhold Limited
1410 Birchmount Road
Scarborough, Ontario MIP 2E7, Canada

Van Nostrand Reinhold Australia Pty. Ltd.
17 Queen Street
Mitcham, Victoria 3132, Australia

Van Nostrand Reinhold Company Limited
Molly Millars Lane
Wokingham, Berkshire, England

16 15 14 13 12 11 10 9 8 7 6 5 4 3 2

Library of Congress Cataloging in Publication Data
O'Connor, Daniel J
101 patented solar energy uses.

Includes index.
1. Solar energy—Patents. I. Title.
TJ810.025 621.47'0272 80-13773
ISBN 0-442-24432-0 (pbk.)

✸ CONTENTS ✸

CONTENTS

PREFACE

This text is designed as a "thinking tool" for homeowners, inventors, scientists, engineers, and students interested in the field of solar energy. It includes over 100 patented inventions that, in some way, have applied solar power to a particular problem.

In the wake of the Three Mile Island nuclear accident and with continuing oil shortages a certainty, the American inventive genius must focus on alternate energy sources. It is hoped that those who use this book will come to appreciate the wide-ranging applications of solar energy as a power and control source, one that can do the same job at a lower cost than many of our present-day machines.

The book illustrates approximately 5 percent of all the patented solar energy inventions and an attempt has been made to give the reader an overview of the solar energy field. Copies of the complete patents may be obtained from your local library, a patent attorney, or by writing the Commissioner of Patents, Washington, D.C.

In presenting the inventions, which are matters of public record, no claim is made as to their particular utility or marketable features. The ideas are not presented as superior or inferior to similar patented systems—they simply show the broad range of solar energy uses. The author has received no remuneration from the inventors.

You should, of course, consult a patent attorney to determine the claimed scope of any inventions shown in the text before making, using, or selling them, at least regarding those for which the 17-year patent term has not yet expired. Also, with regard to any inventions by the reader, a full patent search should be conducted by your patent attorney to determine any possible patent rights that you may have.

As Abraham Lincoln, himself an inventor, said: "The patent system has added the fuel of interest to the fire of genius." It is hoped this book will add to the present "fuel of interest" in the solar energy field.

APPARATUS FOR TRANSMITTING SUNLIGHT TO BASEMENTS OR OTHER STORIES

U.S. PATENT 668, 404

ISSUED TO O. B. H. HANNEBORG, 19 FEBRUARY 1901.

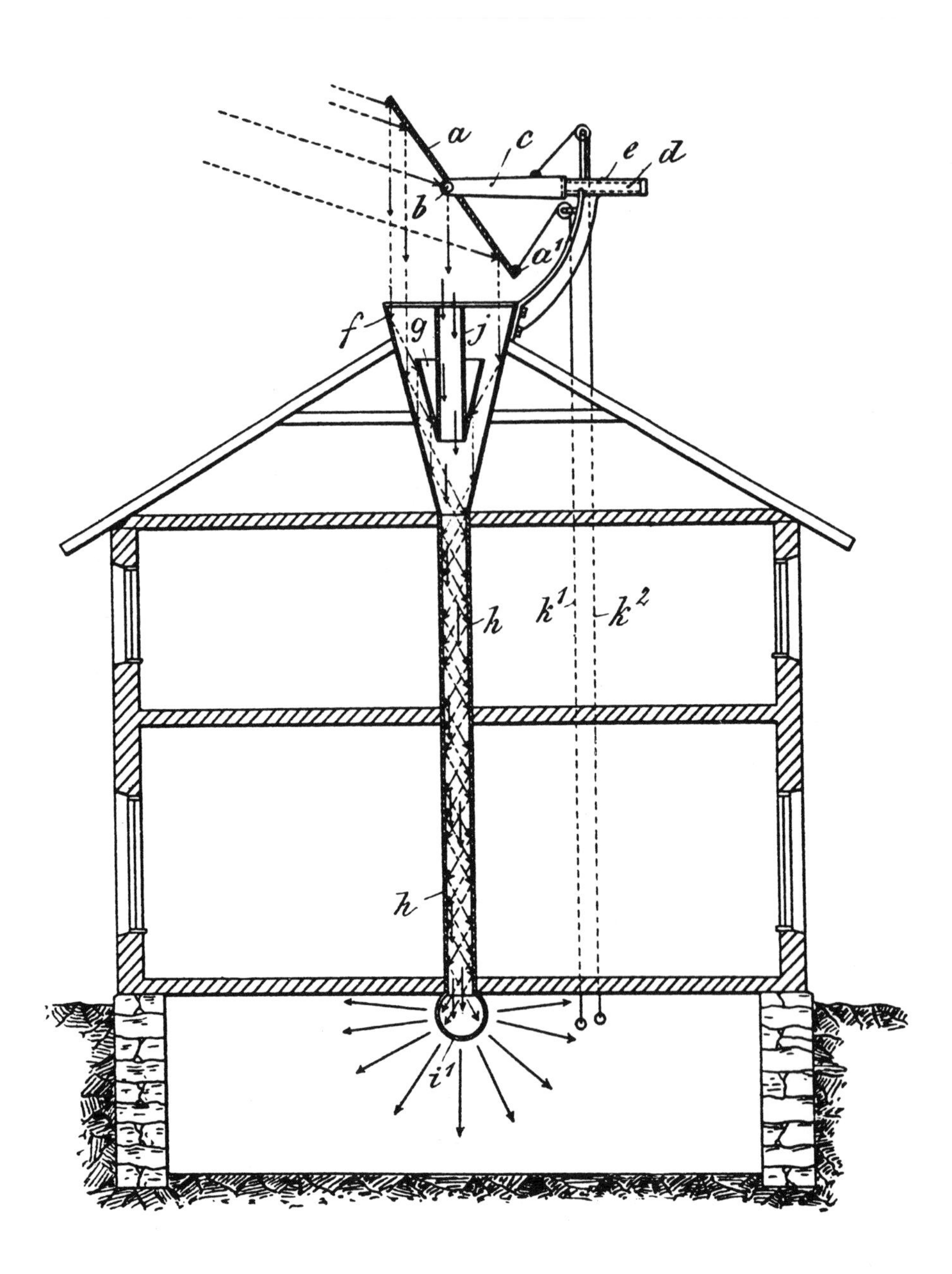

This device utilizes sunlight to illuminate a dark and windowless interior space. Reflector (a) directs light into a light-reflecting funnel (f) and down through tube (h) to the area to be lit. There it is dispersed by globe (i^1). The system could be used to bring nonelectrical illumination to mineshafts, ships, and other areas unreachable by direct sunlight.

APPARATUS FOR USE IN PROPAGATING PLANTS

U.S. PATENT 762, 589

ISSUED TO ROBERT S. LAWRENCE, 14 JUNE 1904.

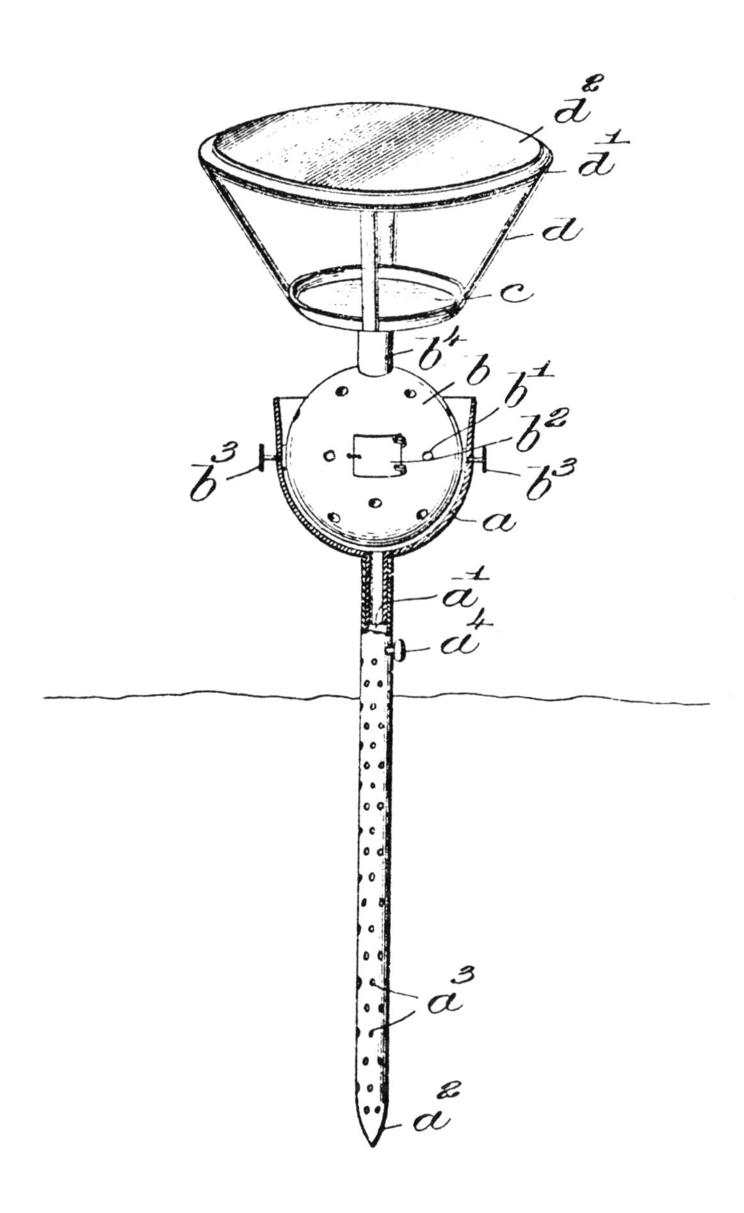

An automatic plant feeder that delivers nutrition directly to plant roots and incidentally warms them frees the owner from time-consuming plant care. Solid plant food is placed in ball (b). Plates (d^2 and c) collect heat from the sun and direct it toward the ball. This liquefies the plant food, which is then carried underground in a measured flow through tube (a^1) to the plant roots.

DRIER

U.S. PATENT 1,073,729

ISSUED TO CHARLES BARNARD, 23 SEPTEMBER 1913.

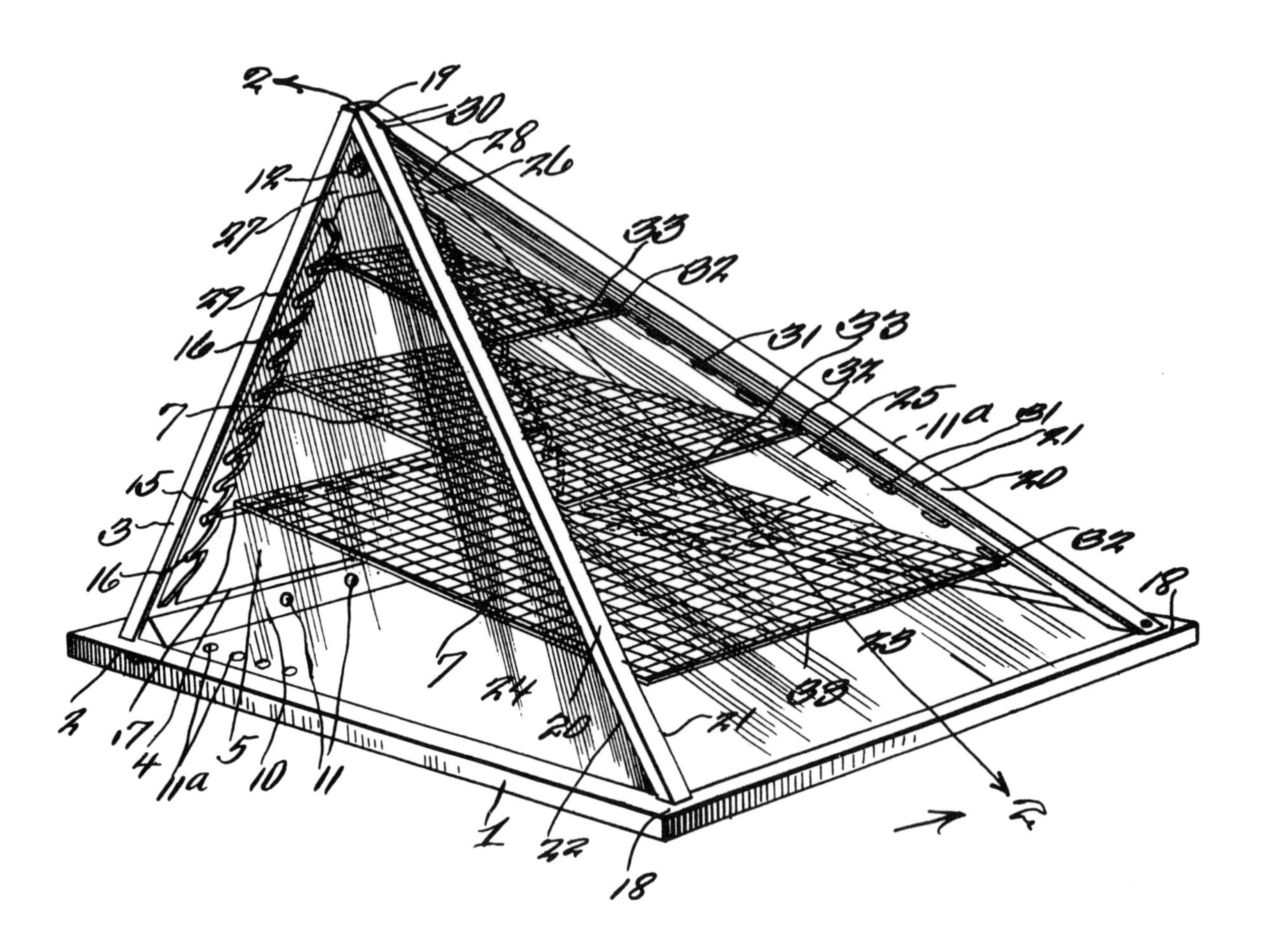

Meats, fruits, and vegetables could be dried and cured with this solar collector. Food to be treated is simply placed on shelves (7) and the structure is sealed to keep in solar heat energy. The pyramid shape of the device captures more sunlight year-round.

LIGHT-DIRECTING BRICK AND WALLS AND BUILDING UTILIZING THE SAME

U.S. PATENT 2,179,863

ISSUED TO THOMAS W. ROLPH, 14 NOVEMBER 1939.

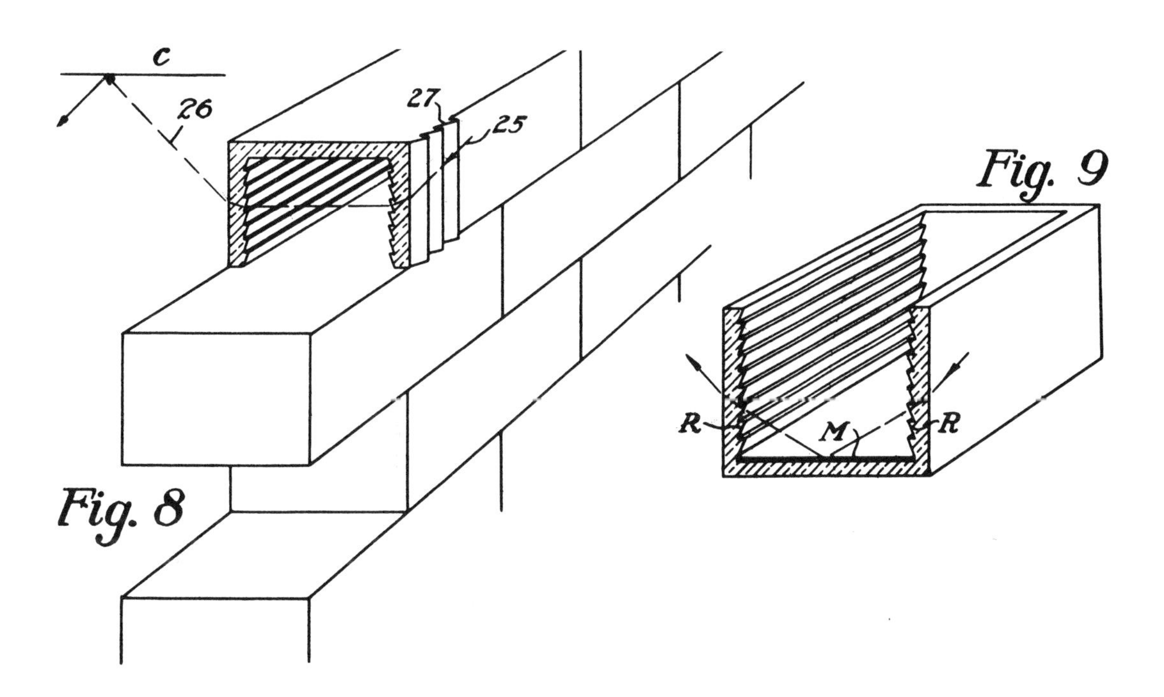

A pleasing indirect lighting effect can be achieved by using shaped building blocks to redirect sunlight. The blocks are supplied with refracting lenses (27) so that sunlight (25), which would normally strike the floor, is bounced upward to the ceiling.

FORMATION OF THERMAL AIR CURRENTS

U.S. PATENT 2,268,320

ISSUED TO ROBERT L. BRANDT, 30 DECEMBER 1941.

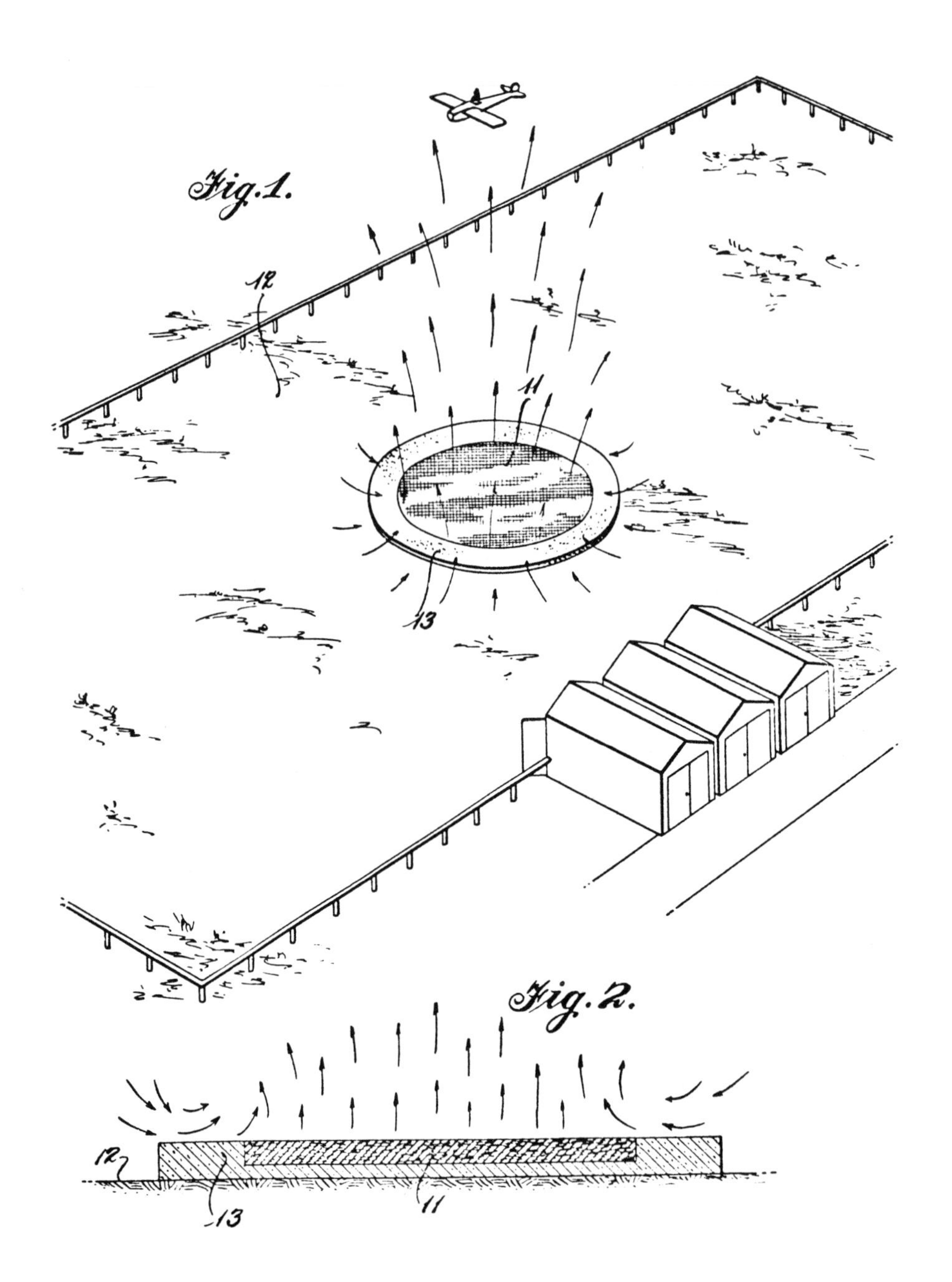

Glider pilots can be lifted to a proper flying altitude by using energy from the sun. A darkened sunlight absorber (11) lying in an open field captures sunlight until the absorber's temperature is raised above that of the surrounding air. With the aid of a white reflector (13), the absorber then releases its heat in the form of an upward air current. All the glider pilot has to do is position his aircraft over the lifting air flow to be propelled to a height where natural wind currents take over.

DISTILLING APPARATUS

U.S. PATENT 2,332,294

ISSUED TO BENJAMIN H. BOHMFALK, 19 OCTOBER 1943.

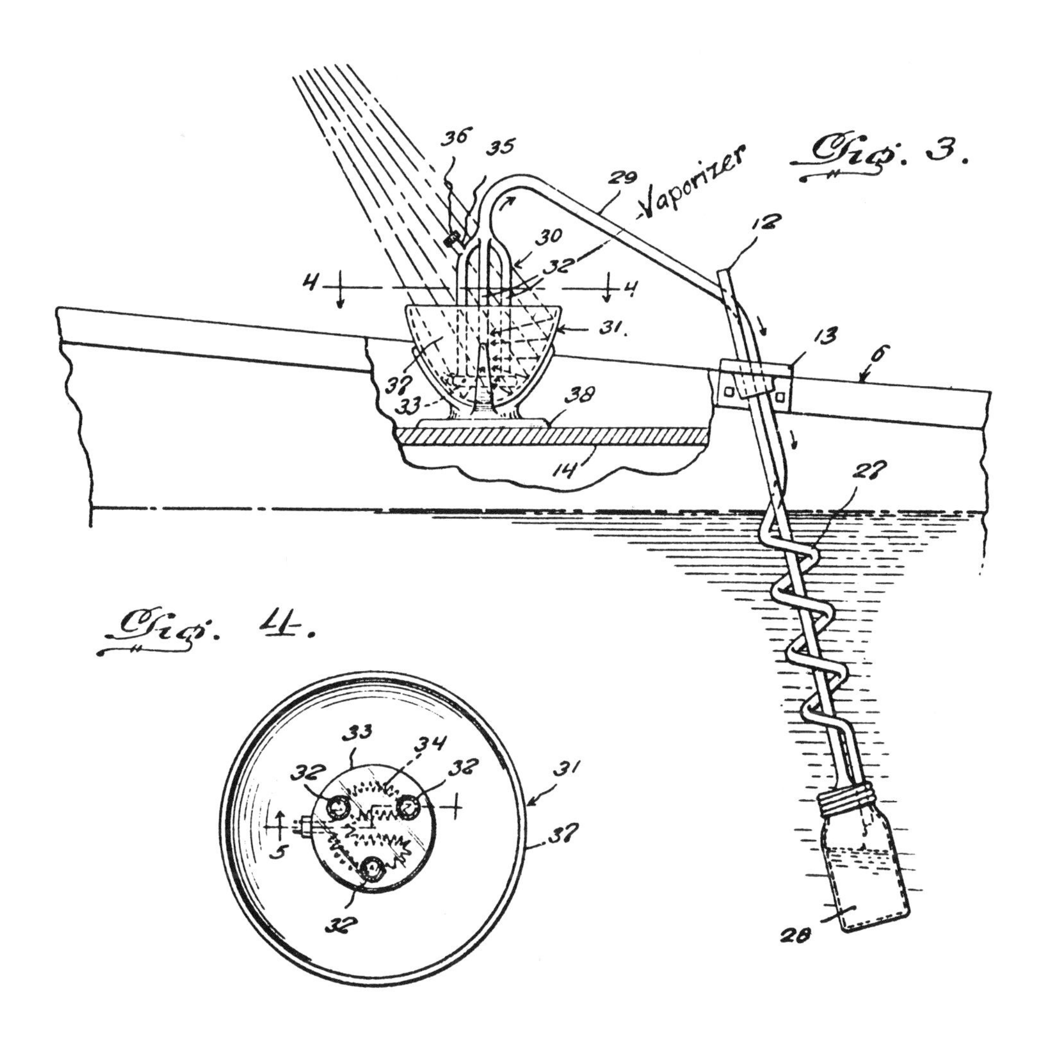

The lives of boaters lost at sea could be saved by using solar energy to obtain drinkable water. Seawater is poured through cap (36) into vaporizer tubes (30), where sunlight reflected onto the tubes causes the water to vaporize. Salt remains in the vaporizer tubes while the salt-free vapor passes down tube (27) into collection bottle (28). Because the tube and bottle are cooled by the seawater in which they are immersed, the vapor condenses as it flows along this route, arriving as fresh potable water. The unit is small enough to be stored and used on a lifeboat.

SOLAR-OPERATED ELECTRIC SWITCH

U.S. PATENT 2,410,421

ISSUED TO JAMES M. BRADY, 5 NOVEMBER 1946.

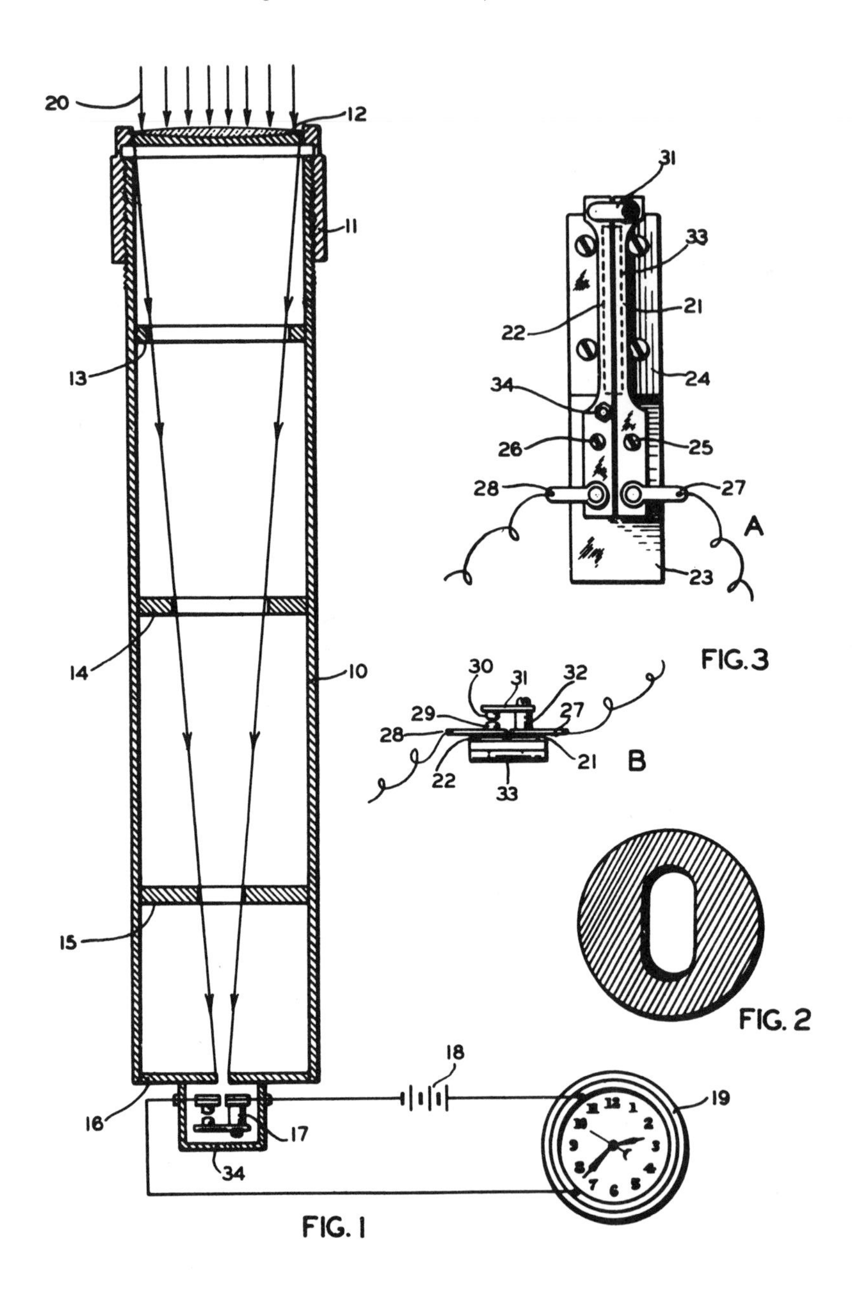

An electric switch can be turned into an automatic timing device by using solar energy. Brady's invention uses the sun to regulate and reset an electric clock at a selected time each day, but many other functions, such as turning on a coffee pot or operating a lamp, can be found for it. The invention operates by concentrating solar heat through lens (12) and tube (10) onto a thermo-responsive switch in circuit (17). When the switch is sufficiently heated, it closes, completing the circuit and turning on the desired device.

HEATING SYSTEM

U.S. PATENT 2,425,797

ISSUED TO ELWYN R. GILLESPIE, 19 AUGUST 1947.

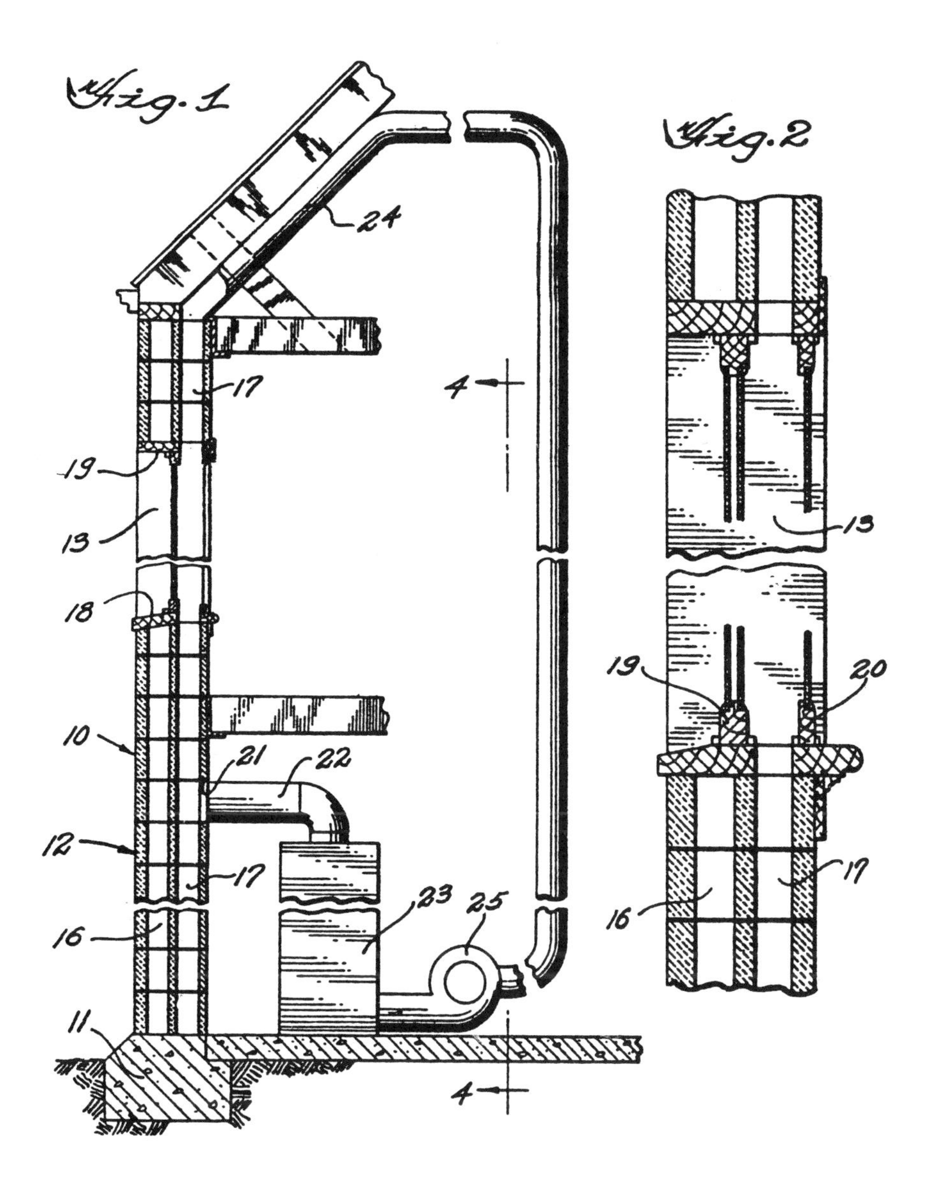

Common building blocks that contain two airways can be used to build a solar heating system. By simply aligning the building blocks to form ducts, a home can become an efficient solar collector without using visible air ducts or other conventional collector structures. In practice, blocks are stacked to form a building wall with a window opening (13) and to provide ducts (16) and (17). The outer duct (16), which is closed by the window sill (18), serves as a layer of air insulation. The inner duct (17) has air forced through it by blower (25). The forced air flows upward between two window panes, picking up solar heat energy that can then be ducted throughout the building.

COCOON SUN SWEAT SUIT

U.S. PATENT 2,478,765

ISSUED TO CHAN J. KIM, 9 AUGUST 1949.

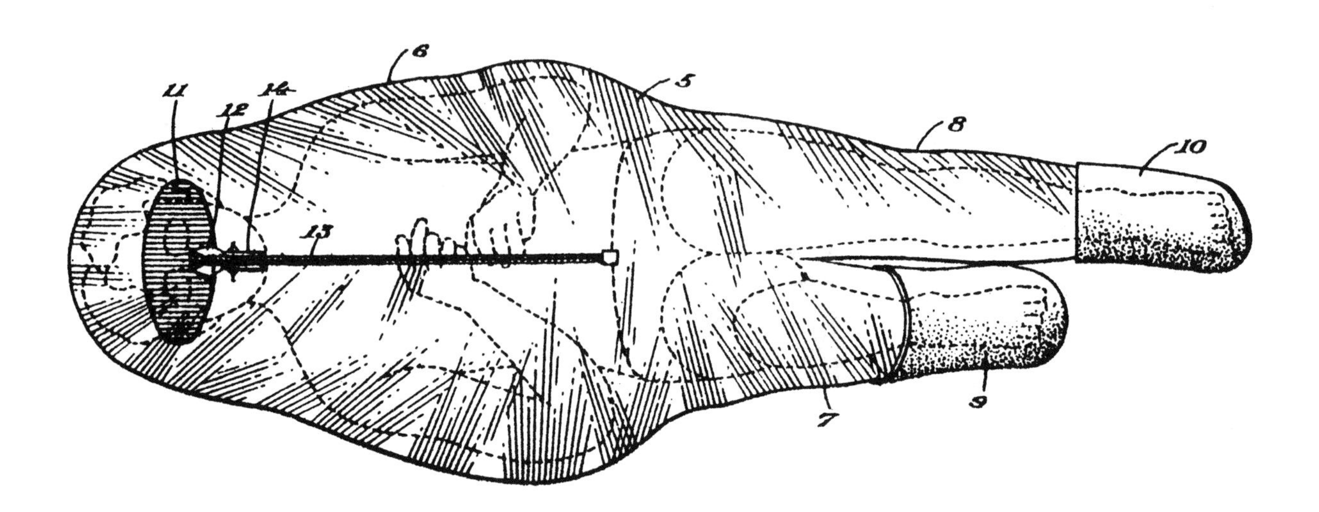

This whimsical item is intended to be a weight-reduction aid, a sort of portable Turkish bath. A transparent membrane (5) collects heat from the sun. An eyeshade (11) and air openings (12) are provided for the comfort and well-being of the wearer.

ROOF OR COVERING

U.S. PATENT 2,489,751

ISSUED TO GEORGE V. CANDLER, JR., 29 NOVEMBER 1949.

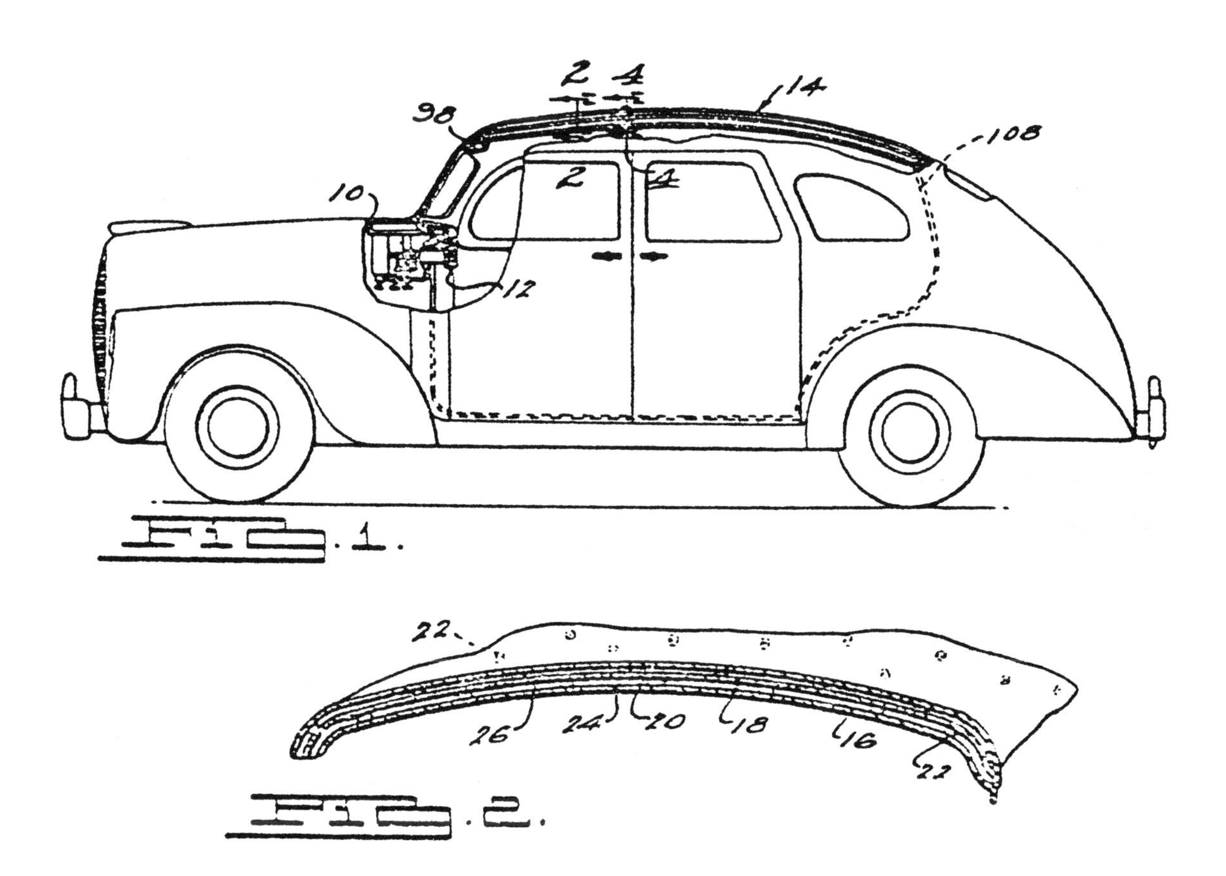

The transparent roofs on some of today's automobiles, used for their pleasant visual effect only, could be helping to cool or heat the passenger area of the car. Candler's invention uses a similar transparent roof (14) with the addition of underlying ducting. In the summer, a light-colored fluid flows through the ducting to reflect heat and perform a cooling function. In wintertime, a dark-colored fluid is used to absorb the solar heat and help warm the car's interior. This type of system could also be adapted for use on the roofs of buildings.

DEHUMIDIFIER SYSTEM

U.S. PATENT 2,566,327

ISSUED TO ROBERT F. HALLOCK, 4 SEPTEMBER 1951

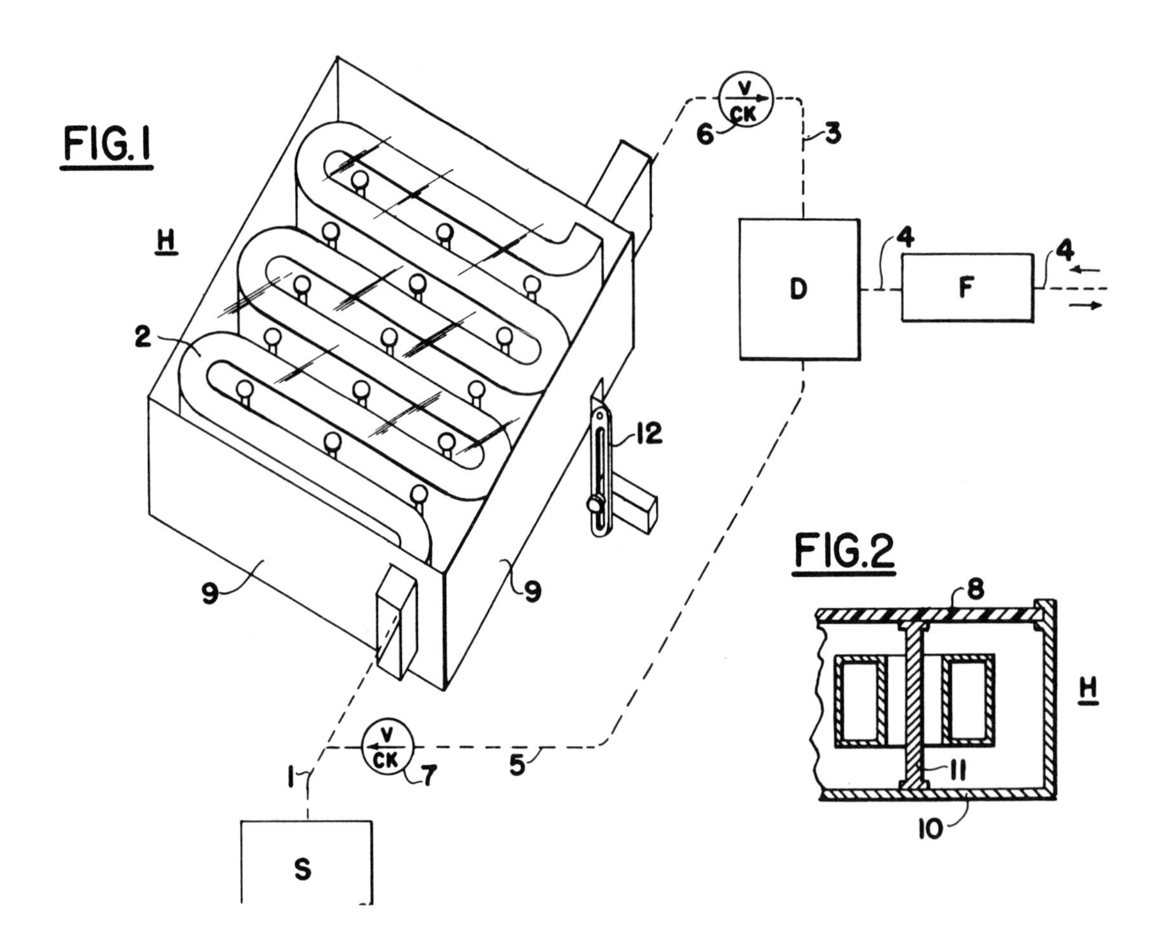

Dehumidifier systems can become more efficient with the help of solar power. Heated air passing through a dehumidifier will be made moisture-free more readily than will cool air passing through the same system. Hallock's device produces heated air by using solar collector (H) with heating tubes (2) inside. As air passes through these tubes, it is heated by the sun. The air is then pumped into dehumidifier (D). From there the dehumidified air flows through filter (F) into the building.

REVERSIBLE SUMMER-WINTER COVER FOR BEEHIVES

U.S. PATENT 2,599,141

ISSUED TO MERRITT I. TAYLOR, 3 JUNE 1952.

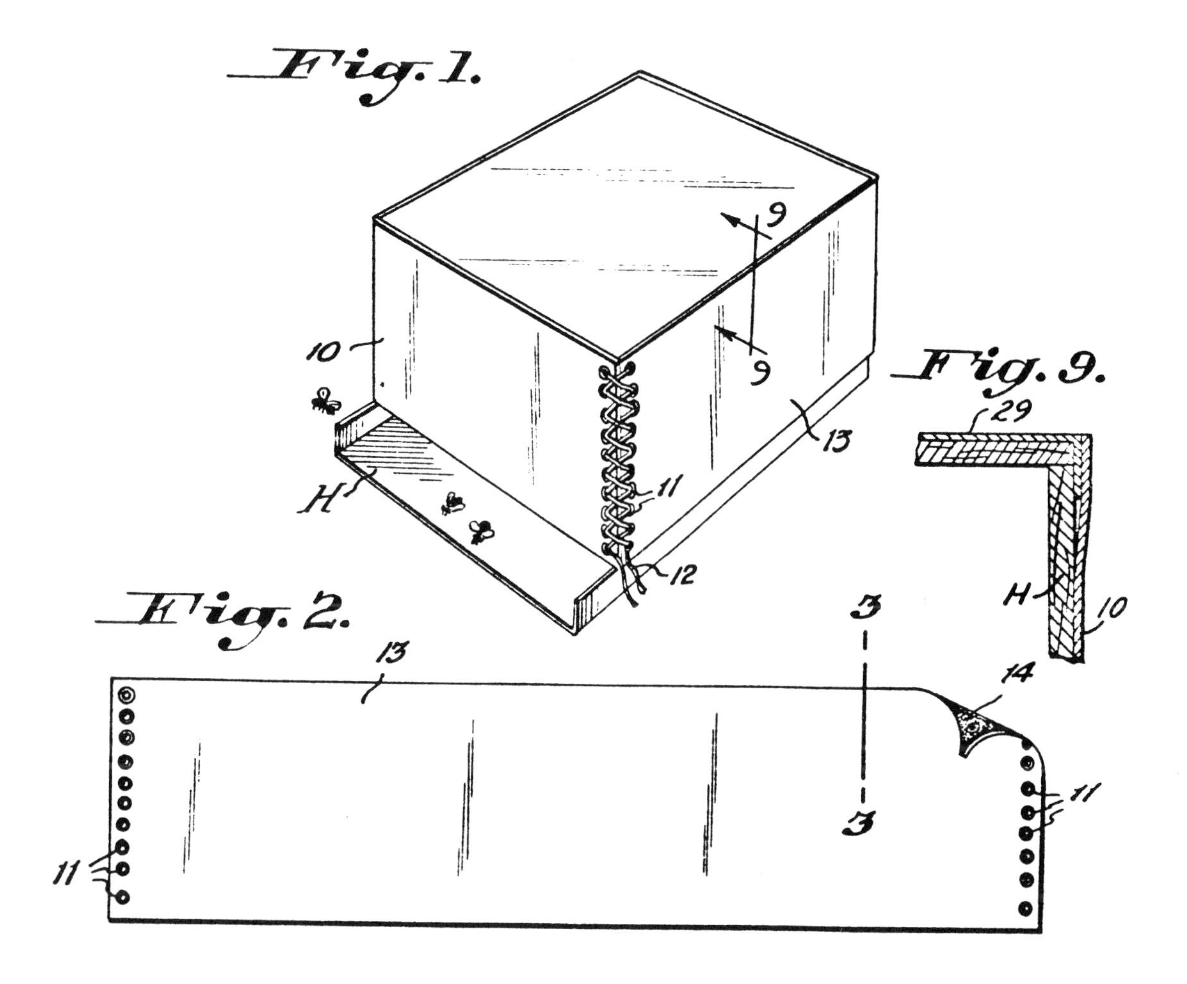

Insects and animals are just as affected by heat and cold as are humans. Taylor's invention utilizes solar energy to keep beehives cool in summer and warm in winter. The outer covering is reversible. On one side (13) is a reflective material used in summer to reduce heat in the hive. In winter, lacing (11) is removed and the cover is turned inside-out to expose the reverse side (14), which is a dark, absorbing material that raises the hive temperature.

SOLAR-ENERGY PUMP

U.S. PATENT 2,688,923

ISSUED TO FILIBERTO A. BONAVENTURA ET AL., 14 SEPTEMBER 1954.

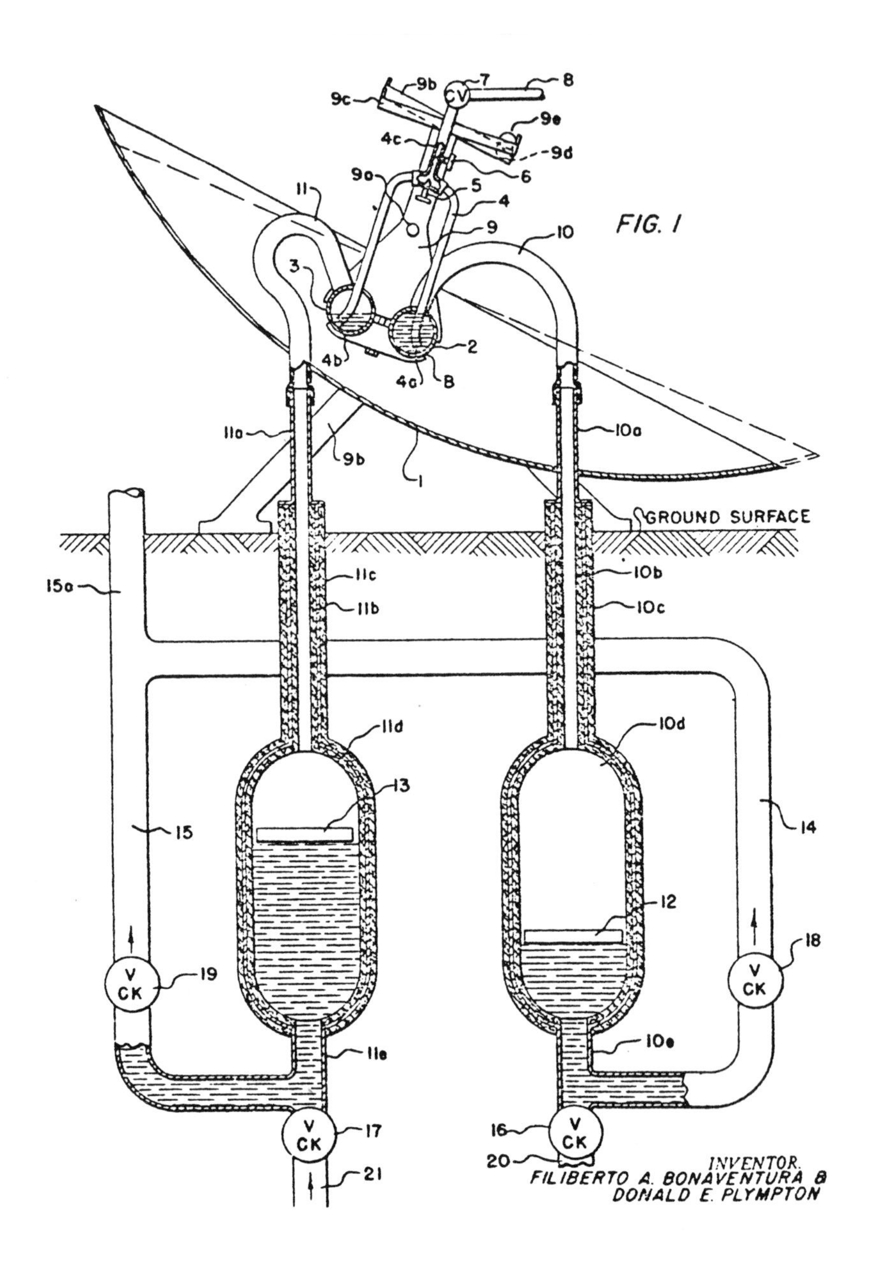

Bonaventura's invention uses a solar-powered pump to bring up underground well water. In operation, liquid flows into chamber (2) where it is boiled into a vapor. This vapor travels through pipe (10a) and forces plate (12) downward, driving the well water through pipes (14) to (15a) and out to the surface. After this action occurs, a pivot about point (9a) changes the focal point of the sun's rays to chamber (3). The same process is repeated on plate (13), forcing more well water to the surface.

MOUTH SAVER

U.S. PATENT 2,809,632

ISSUED TO IVAN VON SOSDY ET AL., 15 OCTOBER 1957.

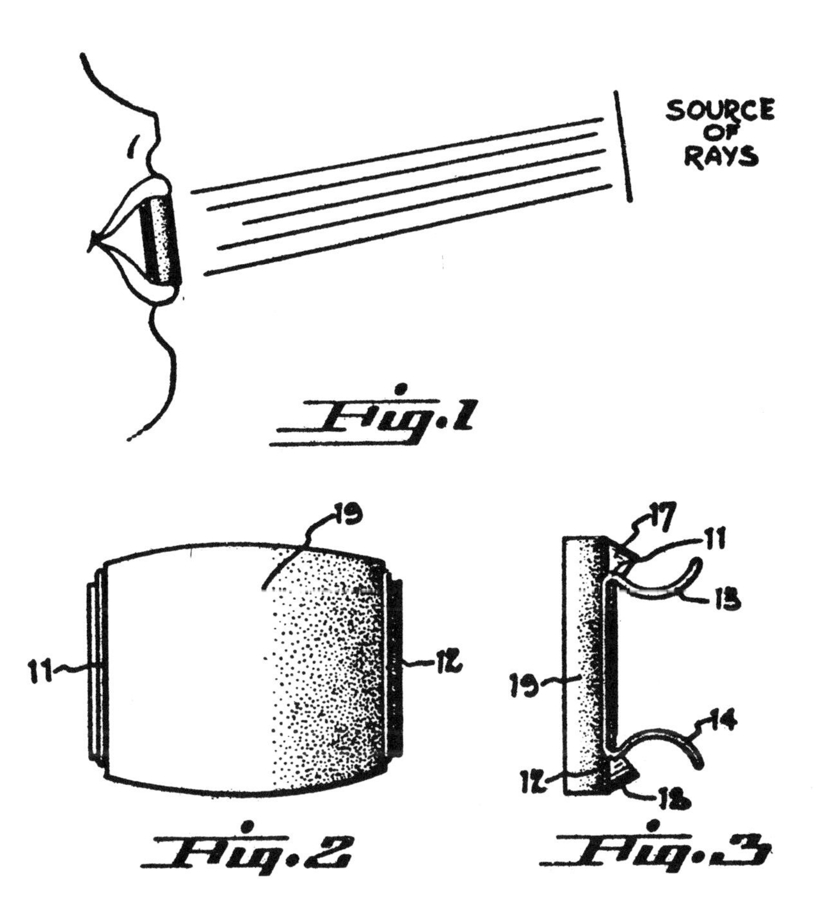

Throughout history many people have believed that the sun has healing powers. This relatively recent invention demonstrates the application of this theory by using sunlight to treat gums and mouth ulcers. In use, the patient places transparent piece (19) with brackets (13) and (14) into his mouth; filtered sunlight is thus focused on the afflicted area.

SOLAR-POWERED TOY BOAT

U.S. PATENT 2,881,558

ISSUED TO KEITH L. BELL, 14 APRIL 1959.

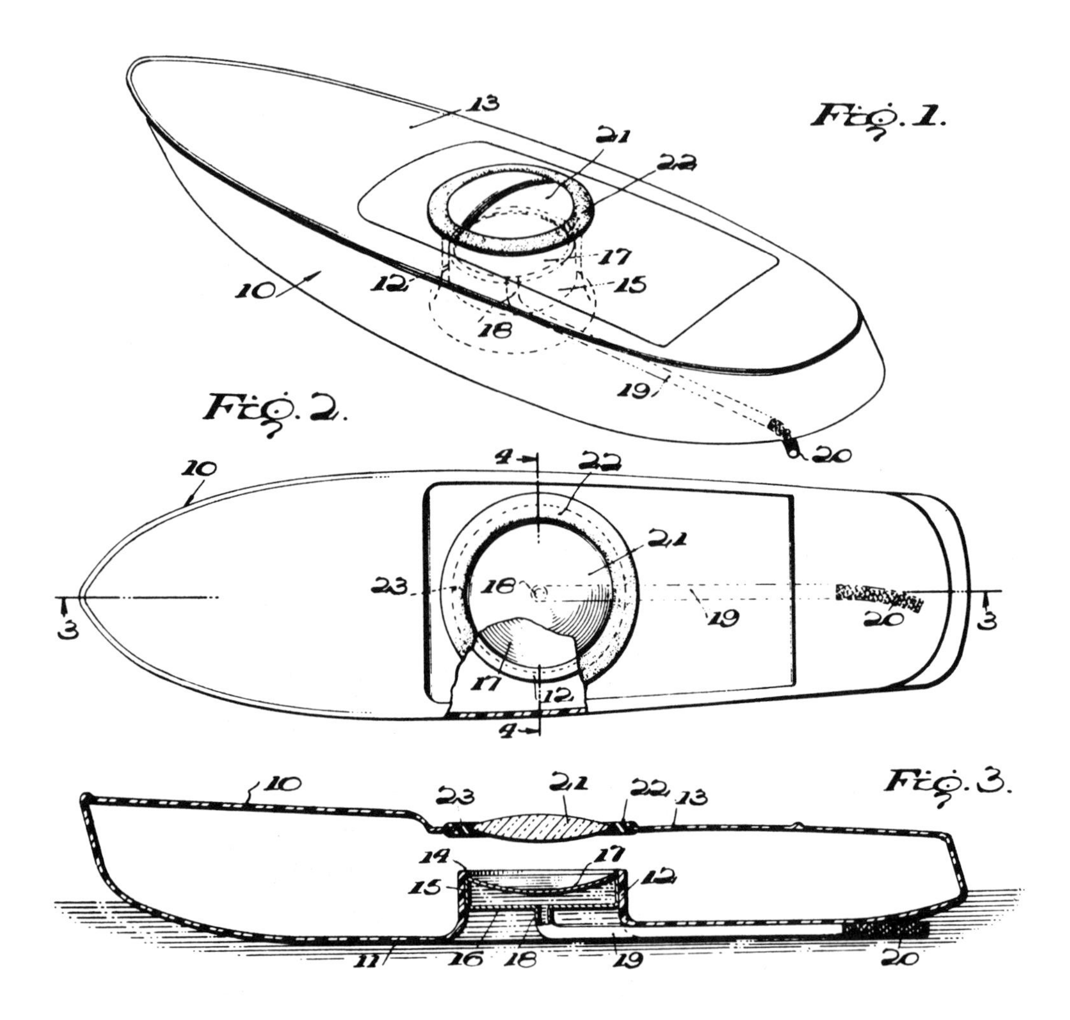

Introducing children to solar energy and its uses can be fun as well as educational with this solar-powered toy boat. Sun rays enter lens (21) and heat plate (17), which is composed of two different metals, one of which expands more rapidly than the other when heated. The sun's heat, therefore, causes the plate (17) to snap down and force water out of exit tube (19). This action propels the boat forward. When plate (17) is in its lower position, it is no longer at the focal point of lens (21) and it cools down. When cold, the plate snaps back up, water is drawn into chamber (15), and the process is begun again.

VEHICLE HAVING A SOLAR STEAM GENERATOR

U.S. PATENT 2,920,710

ISSUED TO GEORGE E. HOWARD, 12 JANUARY 1960.

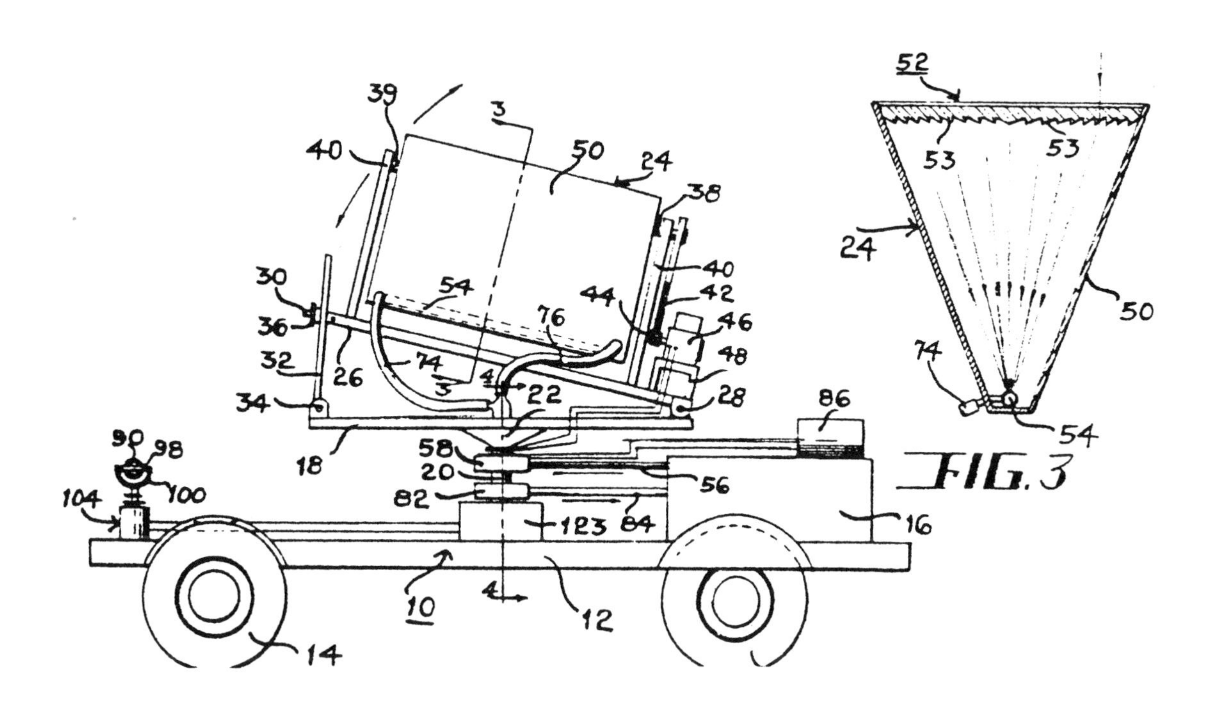

The Stanley Steamer was a popular car at the turn of the century. This invention shows how steam generated by a solar collector can drive an automobile. In operation, a Fresnel lens (53) and solar concentrator are employed in container (50) to generate steam, which is fed through lines (76) and (84) to a steam engine (16). The power produced in the steam engine turns the wheels of the car. After condensation in the steam engine, the fluid is returned to the solar collector via lines (56) and (74). The container (50) is continuously turned to face the sun by electric motor (46), which also derives its power from steam engine (16) via electric generator (86).

SOLAR THERMOSTAT CONTROL UNIT

U.S. PATENT 2,928,606

ISSUED TO WILLIN C. LEE, 15 MARCH 1960.

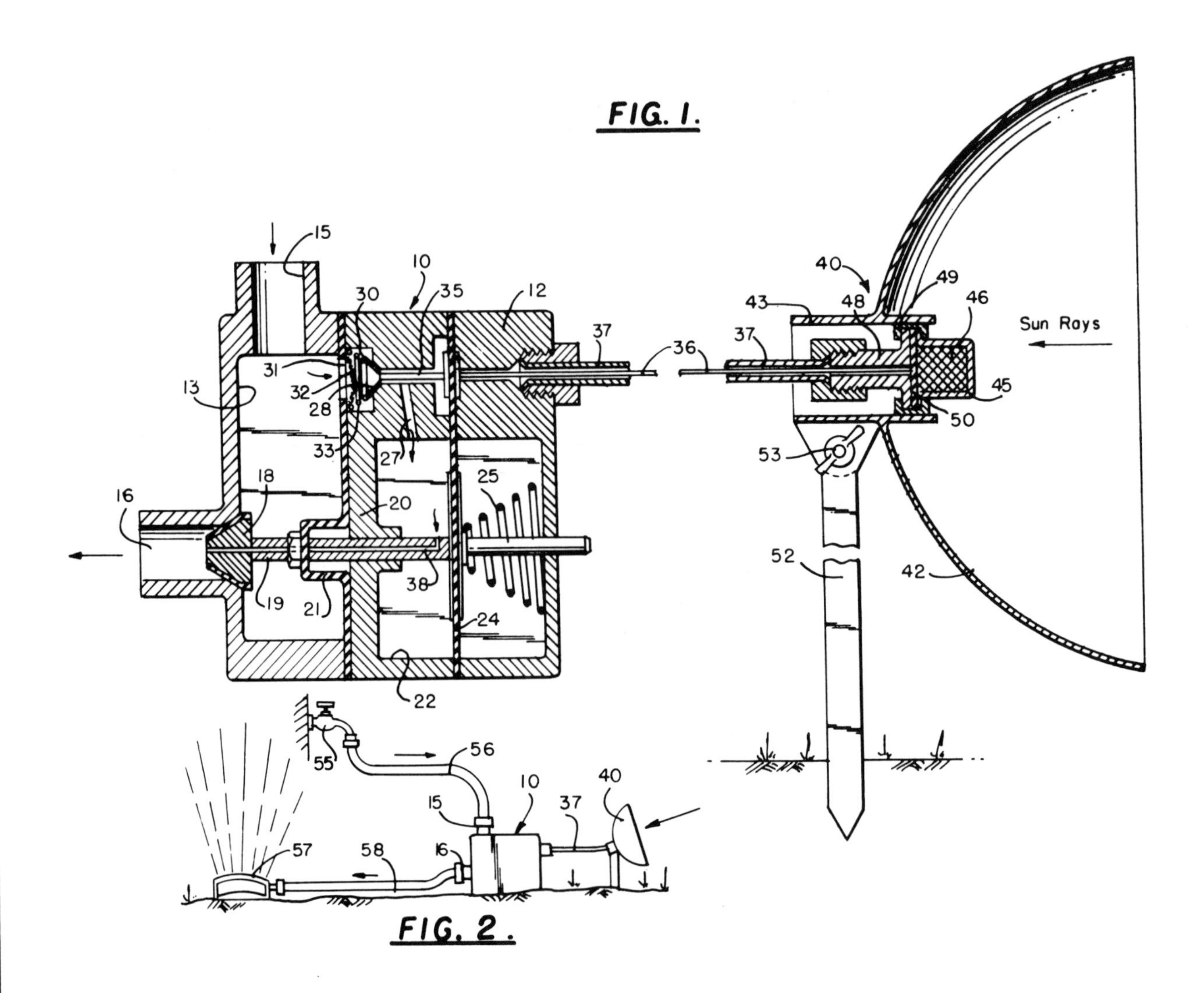

A sprinkler system for a lawn or garden, or an irrigation system in a remote location, can be turned on automatically by using solar power. Reflector (42) directs solar rays to absorber (46). Heat expansion of the absorber acts through rod (36) to open small valve (30), allowing pressurized water to flow around it and through small passage (27) to put pressure on diaphragm (24). The pressure moves the diaphragm to the right and opens valve (18). Once the valve is opened, water from the main supply flows into the sprinkler or irrigator.

CIRCULATING AND HEATING MEANS FOR BUILDING

U.S. PATENT 2,931, 578

ISSUED TO DEAN L. THOMPSON, 5 APRIL 1960.

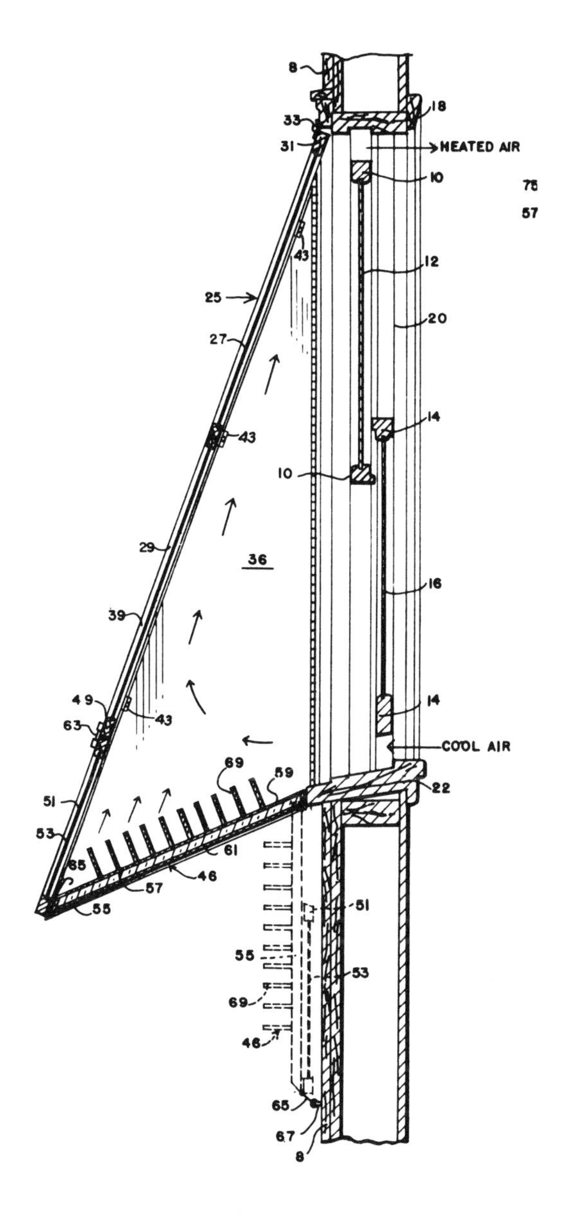

A home window can easily be modified to become a more efficient solar collector. When solar energy collection is desired, the upper glass pane (27) is swung outward and latched to a lower glass pane (53) at point (63). Lower sunlight absorbers (69) and (59) give off their collected heat to cool air flowing in at (14). The heated air then flows over upper window (12) and into the home. During summer months, when maximum solar collection is not desired, latch (63) is unhooked and the absorber (69) is folded down against the building. Upper pane (27) then swings in to lie flush with the structure.

SOLAR-ACTUATED UMBRELLA-RAISING MECHANISM

U.S. PATENT 2,960,094

ISSUED TO SAMUEL N. SMALL, 15 NOVEMBER 1960.

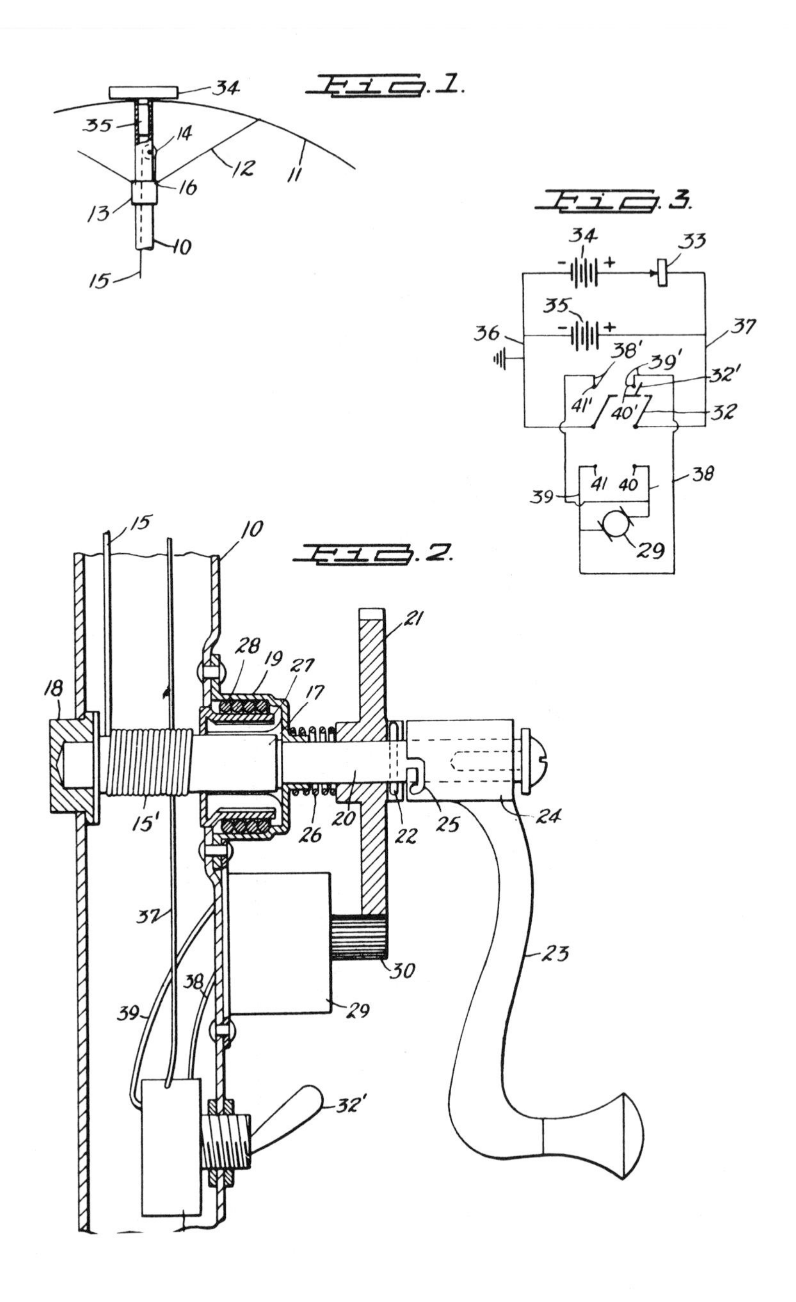

Solar-powered umbrellas can be used by hotels at beach and pool areas, where it may be necessary to raise a large number of them quickly. A solar-powered battery (34) is placed above the umbrella (11) to collect sunlight and store an electric charge. When switch (32[1]) is thrown, power is supplied from the battery to the mechanical crank to automatically raise the umbrella.

INFLATABLE CURVED MIRROR

U.S. PATENT 3,054,328

ISSUED TO JOHN RODGERS, 18 SEPTEMBER 1962.

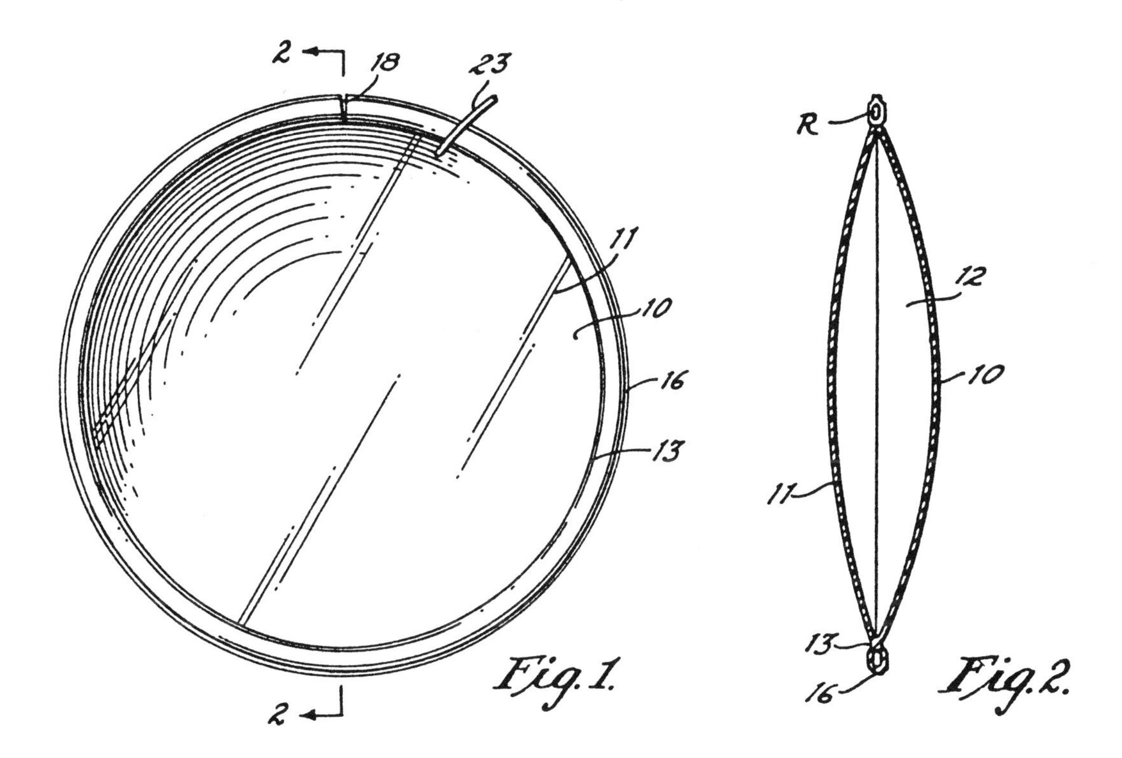

The particular shape of a sun-reflecting mirror is often critical in a solar-energy collecting system. In this device, mirrors (10) and (11) have a shape that can vary depending upon the amount of air pumped into the chamber (12) by hoses (16) and (R).

THERMODYNAMIC RECIPROCATING APPARATUS

U.S. PATENT 3,117,414

ISSUED TO FARRINGTON DANIELS ET AL., 14 JANUARY 1964.

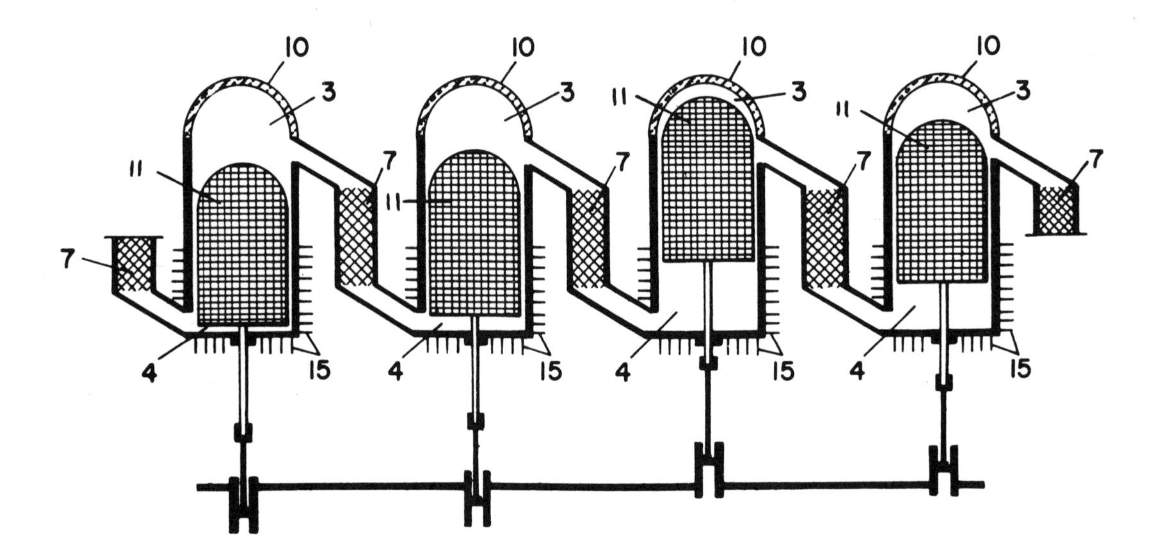

An engine can be powered by sunlight instead of by expensive fossil fuels. Upper air space (3) is heated by solar rays passing through glass (10). The resulting air expansion forces piston (11) downward to drive the lower crankshaft. The heated air then flows down through tube (7) into the lower cylinder area (4), where it drives the next piston (11) upward and is cooled by means of fins (15).

OFF-AXIS FOCUSED SOLAR HEATER

U.S. PATENT 3,179,105

ISSUED TO GERALD FALBEL, 20 APRIL 1965.

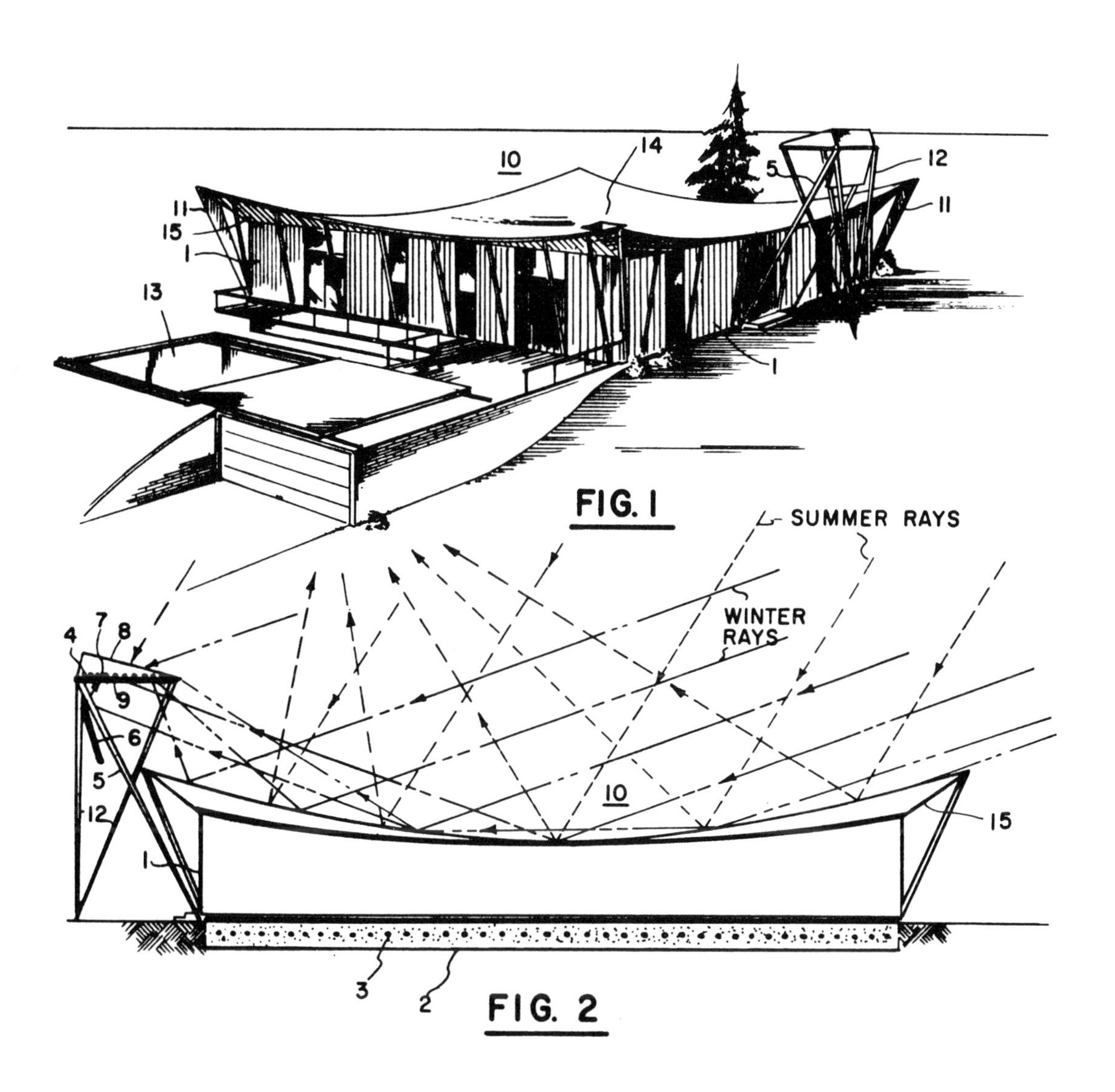

The sun is at a higher angle during the summer than in the winter. Falbel's system takes advantage of this fact by shaping roof (14) and placing collector (8) so that in summer solar rays are reflected away from the collector. In winter, sun rays at a lower angle are reflected into the collector (8) to heat water, which can then be stored in lower concrete slab (2). Thus, no system changes are required for year-round operation.

EAVES TROUGH WITH RADIATION-ABSORBING ATTACHMENT

U.S. PATENT 3,207,211

ISSUED TO IRVING J. WINTERFELDT, 21 SEPTEMBER 1965.

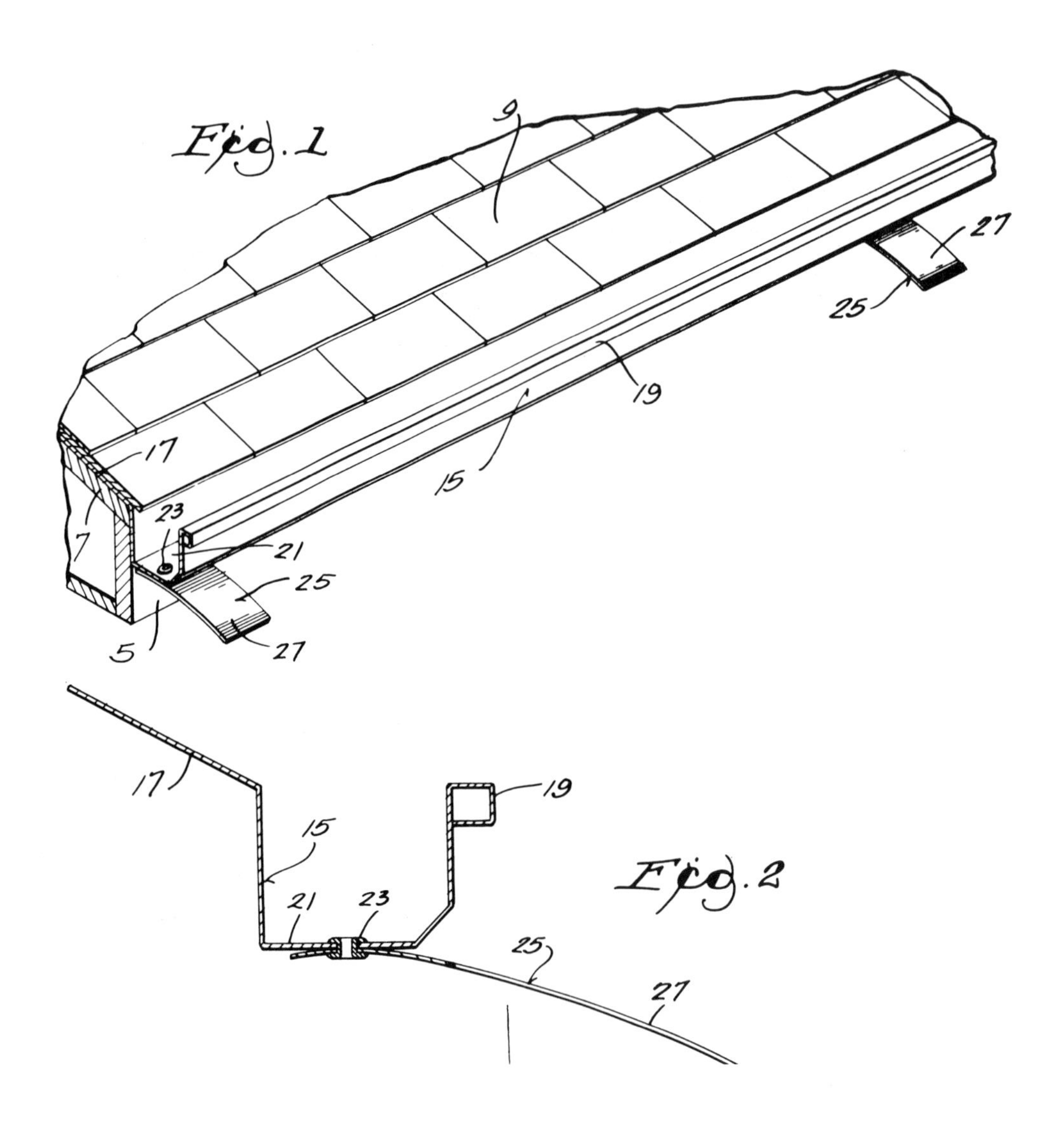

Homeowners in colder climates often have the recurring problem of ice forming in gutter troughs. The weight of ice build-up can cause gutter damage or rot and can result in water leakage through the roof of a structure. The Winterfeldt invention uses solar energy to help solve this problem. Sun-absorbing (i.e., black-coated) plates (25) and (27) are attached to a gutter several feet apart. The plates catch more available sunlight than vertical gutter pieces (15) and transfer the sun's heat energy to the area where it is needed to help prevent ice build-up.

SOLAR BALLOON OR AEROSTAT

U.S. PATENT 3,220,671

ISSUED TO LELAND E. ASHMAN ET AL., 30 NOVEMBER 1965.

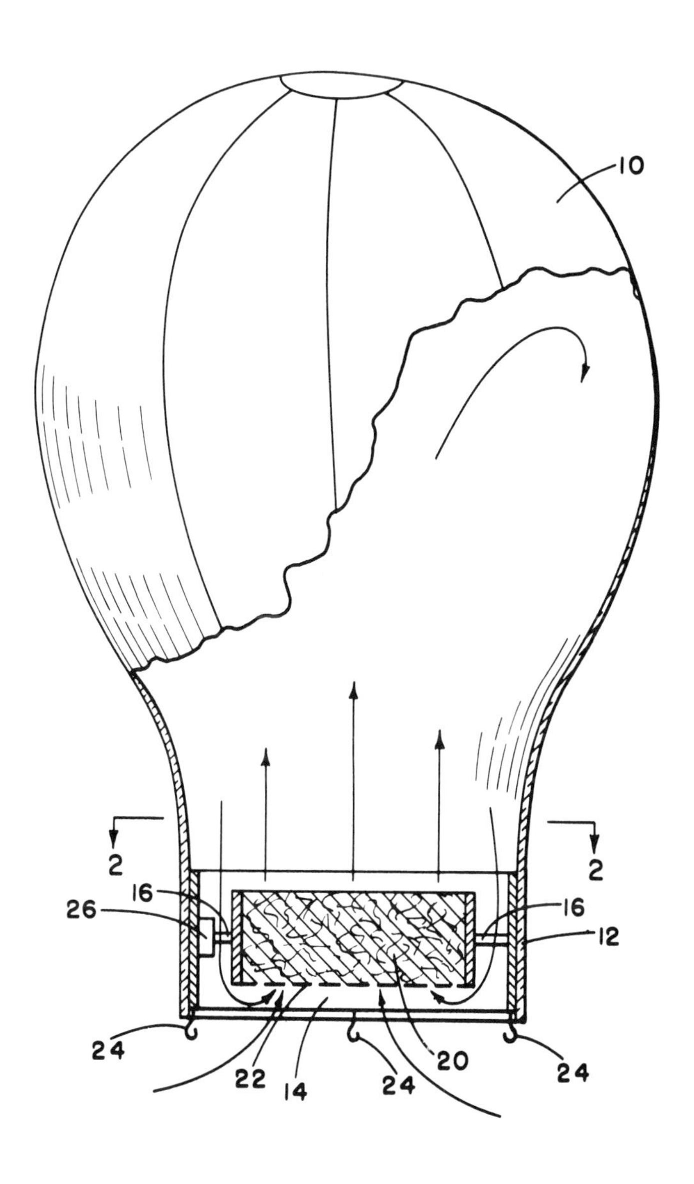

Weather or toy balloons can be propelled by solar energy. In the Ashman device, transparent balloon (10) has dark sun-absorbing fibers (20) attached to its lower end. In operation, sunlight is captured by the fibers and the resulting heat released from them drives the balloon upward.

THERMOELECTRIC-POWERED SATELLITE

U.S. PATENT 3,225,208

ISSUED TO RAYMOND WOLFE, 21 DECEMBER 1965.

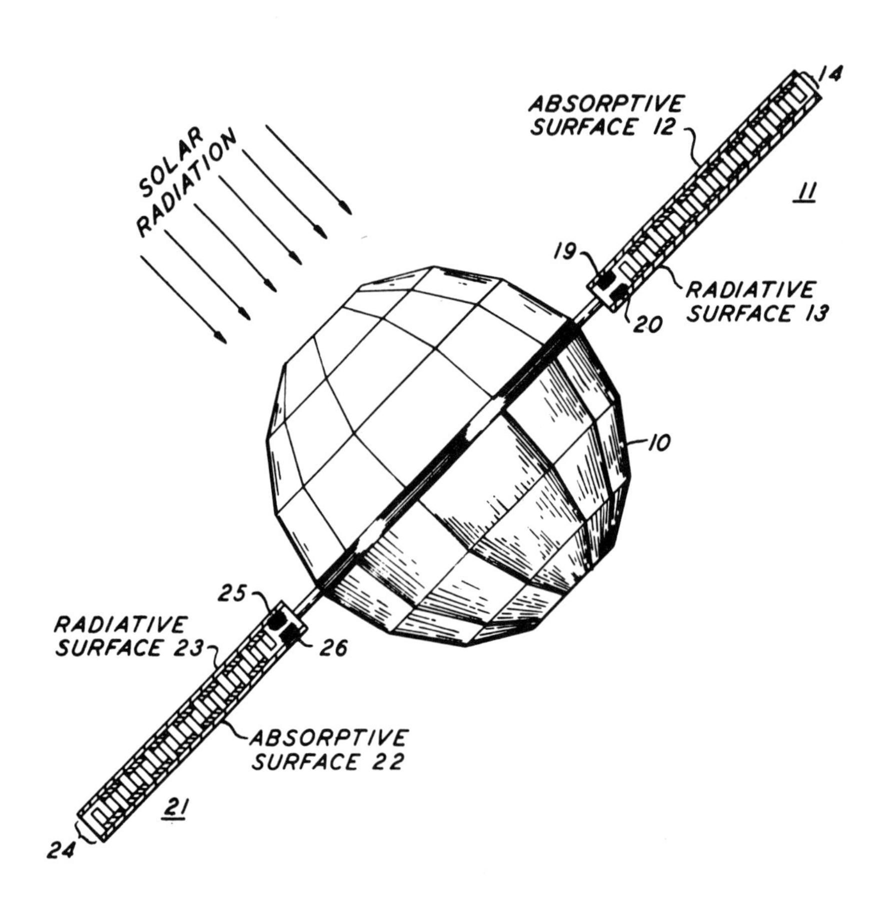

Since there are no clouds in outer space to block the sun, capturing solar energy is relatively easy for satellites. In Wolfe's satellite design, panels (11) and (21) absorb solar radiation and convert it into an electric current. This current is then used to supply electric power to computer and other systems within the satellite (10).

26

APPARATUS FOR CURING TOBACCO

U.S. PATENT 3,231,986

ISSUED TO R. D. TOUTON, 1 FEBRUARY 1966.

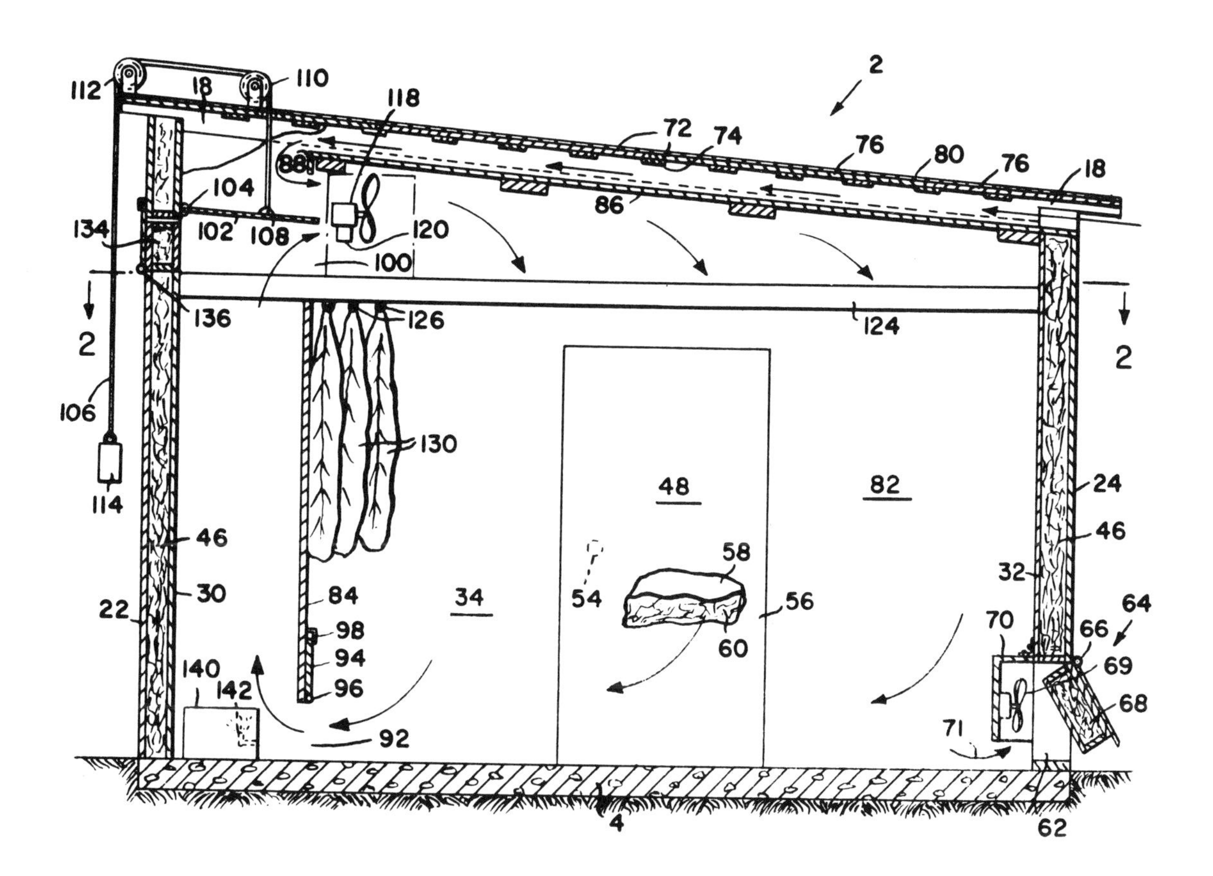

The curing of plants such as tobacco, which requires heat during the process, can be achieved by using solar energy. In the Touton system, air is warmed by the sun in upper duct (86) and circulated by fan (118) over tobacco leaves (130). The hot air can be recirculated through lower passage (92) to further dry the tobacco or can be vented to the outside by lower fan (69) when sufficient drying has occurred.

THERMOSYPHON SOLAR HEAT CELL

U.S. PATENT 3,254,644

ISSUED TO FRED G. THANNHAUSER, 7 JUNE 1966.

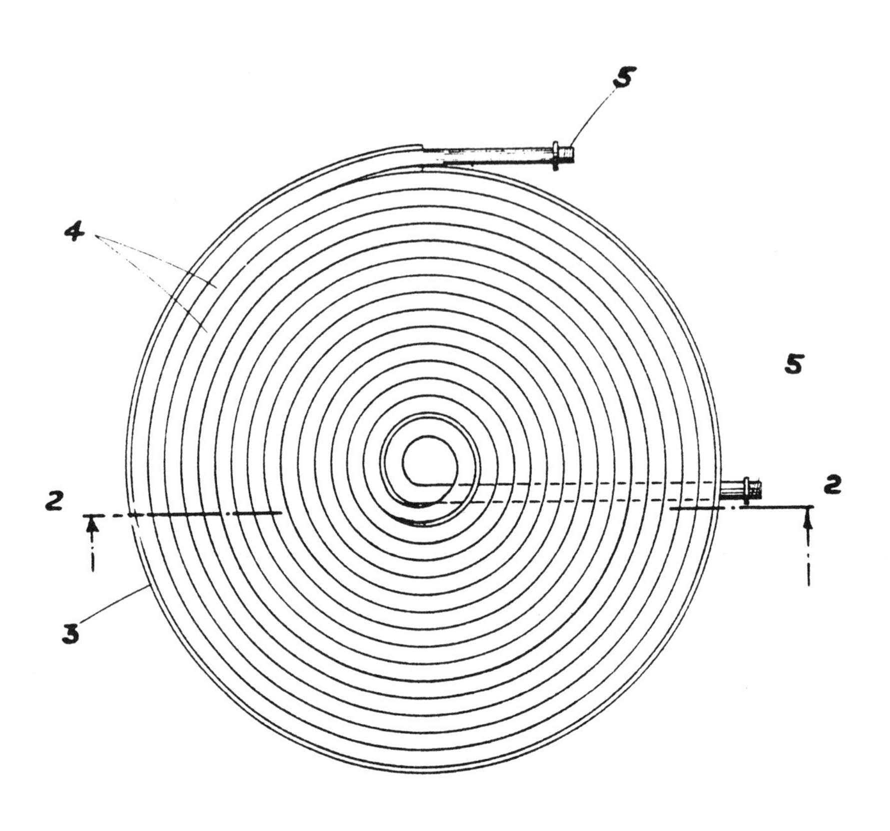

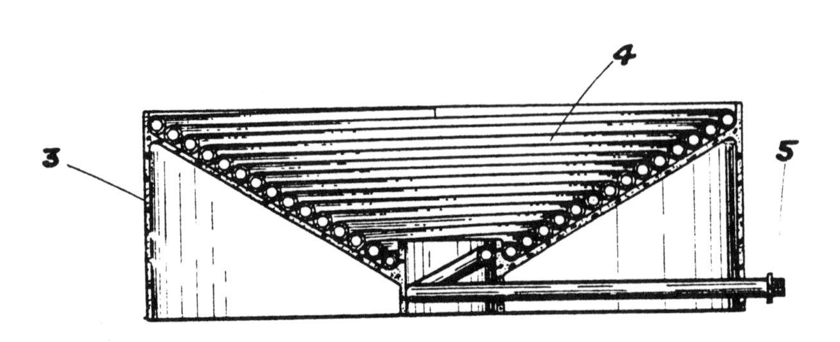

This solar collector is shaped to act as a low-cost pump. In operation, fluid entering at lower port (5) is warmed by sunlight and tends to rise upward, assisted by the gradual slope in the spiral tubing (4). The hot fluid then exits at upper port (5) for use as needed. In heat-transfer terms, the process is known as the thermosyphon effect.

MEANS FOR HEATING AND COOLING A STRUCTURE

U.S. PATENT 3,262,493

ISSUED TO DAVID E. HERVEY, 26 JULY 1966.

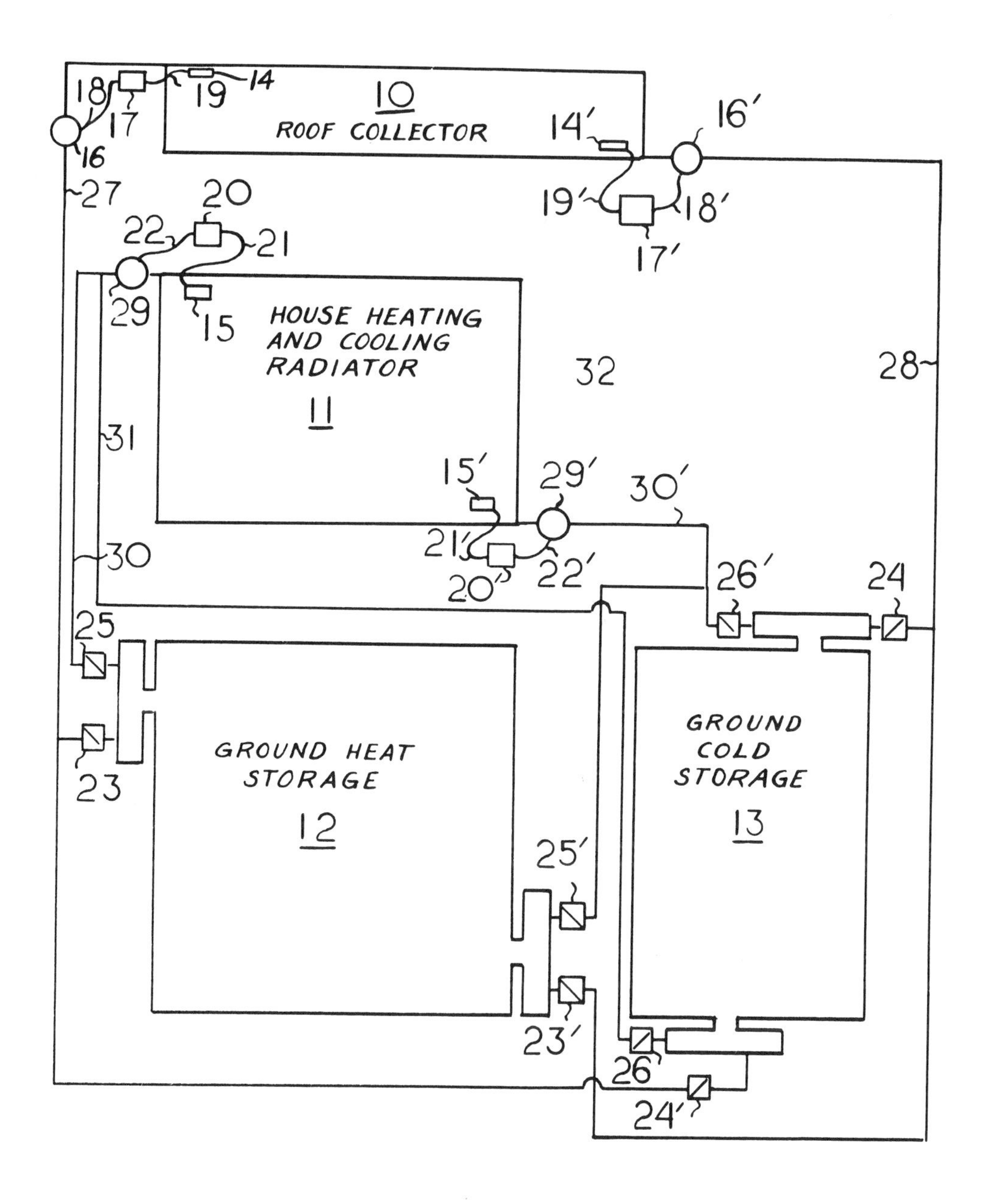

Underground storage tanks allow a home to be both heated and cooled by a single solar collector. Solar heat energy collected on hot summer days is stored in underground tanks (12) for use in winter through radiator (11). While the hot fluid is warming a home, fluid from a cold storage tank (13) is pumped to the roof collector (10) to be cooled for use in the same radiator in summer months.

SOLAR STILL APPARATUS FOR EXTRACTING WATER FROM SOIL

U.S. PATENT 3,290,230

ISSUED TO GUNJI KOBAYASHI, 6 DECEMBER 1966.

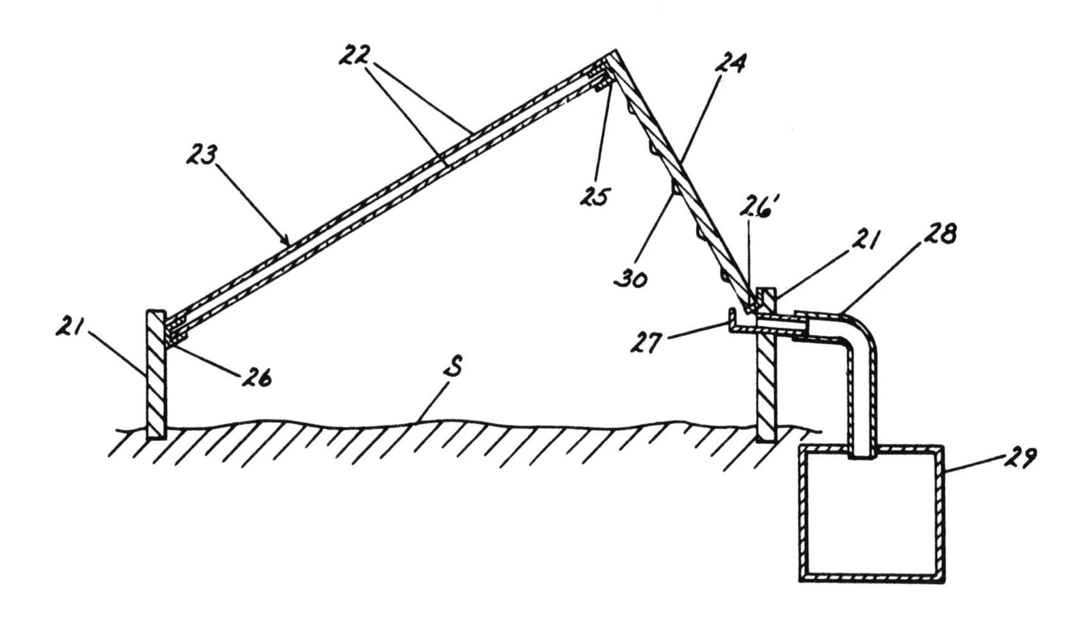

Even in a desert, moisture is present near the ground surface and in the air and it can be collected by using the sun. In this device, a sealed container is placed over the ground; sunlight passing through glass (22) heats the air inside. Plate (24) is relatively cool because of heat transfer from outside the container. Thus, when the hot inside air contacts the cooler plate (24), condensation drops (30) form and are drained into chamber (29).

COOLING UNIT FOR A HAT

U.S. PATENT 3,353,191

ISSUED TO HAROLD W. DAHLY, 21 NOVEMBER 1967.

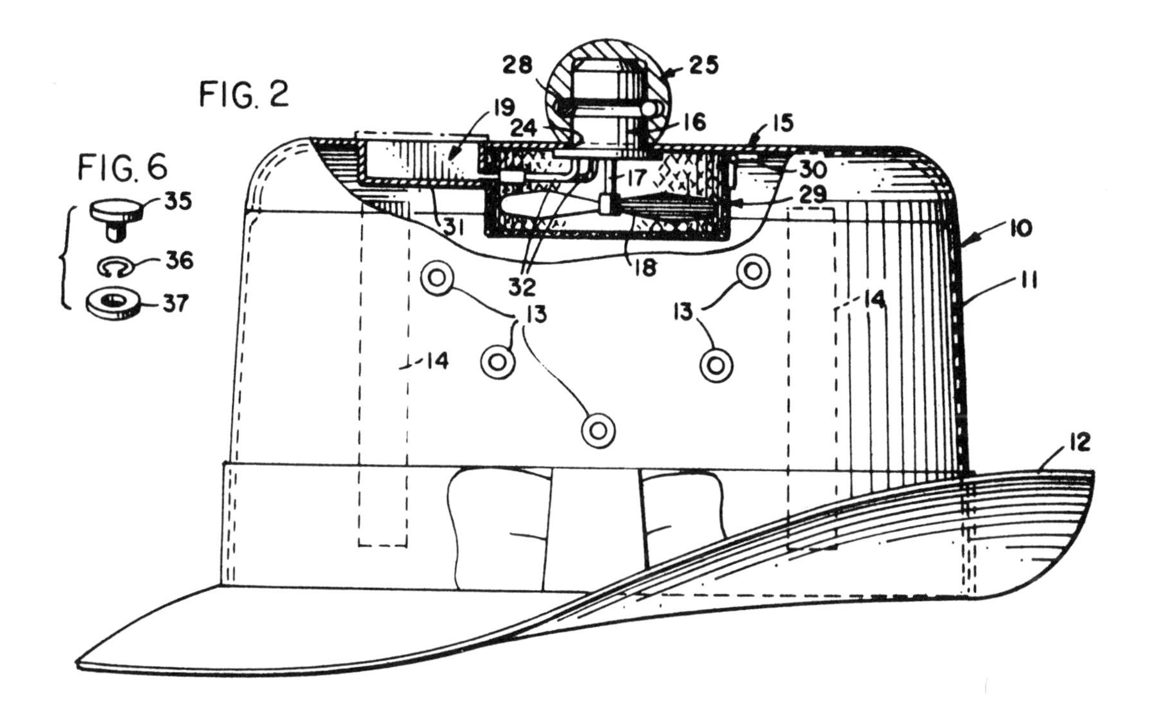

Dahly must have had the term "keeping a cool head" in mind when he invented this solar-powered cooling unit for a hat. In his device, sunlight strikes solar cells (19) and an electric current is produced. This current is used to drive a small electric motor (16), which, in turn, rotates fan (18). Cooling air is then forced through screen (29) to cool the hat wearer's head. Though this is an eccentric device, the system could be put to more practical use where a small cooling fan is needed.

SOLAR AIR-MOVING SYSTEM

U.S. PATENT 3,436,908

ISSUED TO VUKASIN VAN DELIC, 8 APRIL 1969.

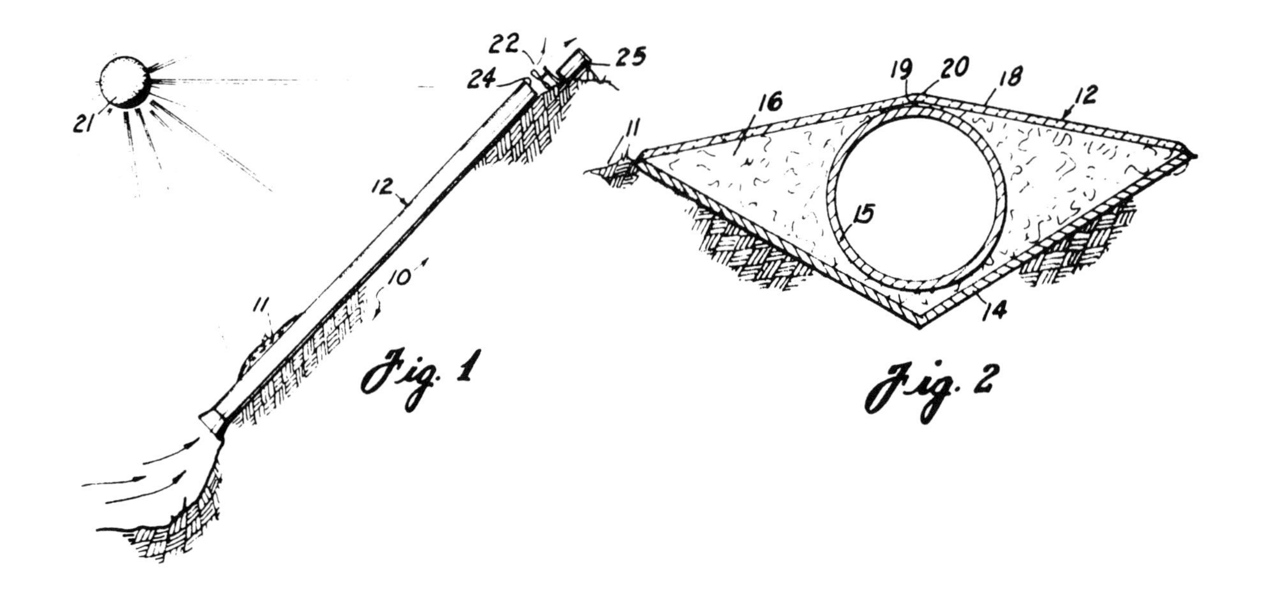

Free electricity can be produced by solar collectors located in mountainous areas. This system places large tubes (12) on a mountainside facing the sun. Air inside the tubes is heated by the sun causing it to rise at a high speed and in large quantities. This rapidly moving air is then used at the upper end of the tube to drive a fan (22) connected to an electrical generator (25). A heat-conducting substance (16) may be included between outer tube (12) and inner tube (15) to enhance the air-heating effect.

SOLAR THERMAL BLANKET

U.S. PATENT 3,453,666

ISSUED TO HENRY M. HEDGES, 8 JULY 1969.

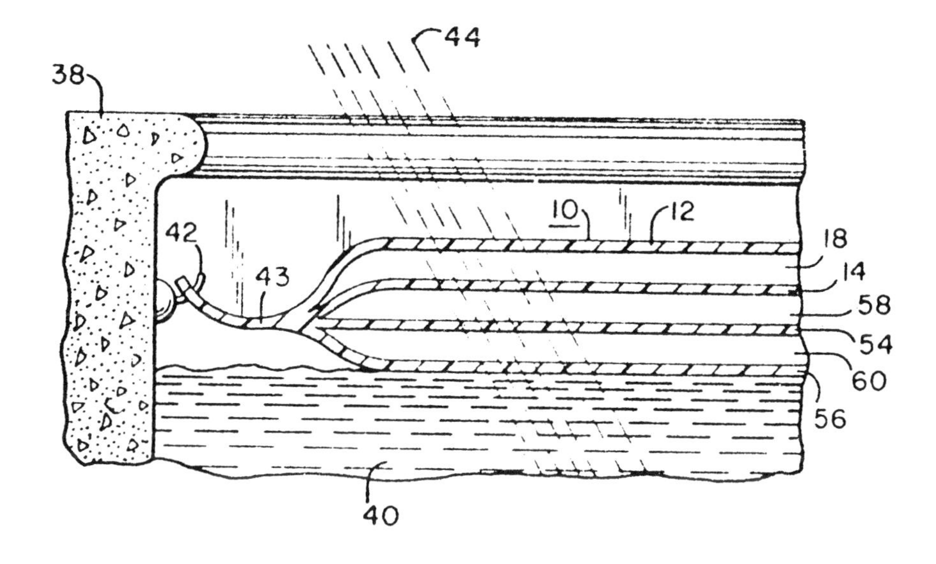

Swimming pools can be warmed by the sun instead of by costly electric heaters. This device uses a clear plastic cover with insulating air layers to capture and retain the sun's heat in pool water (40). In winter, the plastic cover keeps dirt particles out of the pool. Sun-warmed air in passages (18), (58) and (60) could be pumped out to heat a home.

GOLF-BALL WARMER

U.S. PATENT 3,497,676

ISSUED TO KENNETH W. GRAVATT, 24 FEBRUARY 1970.

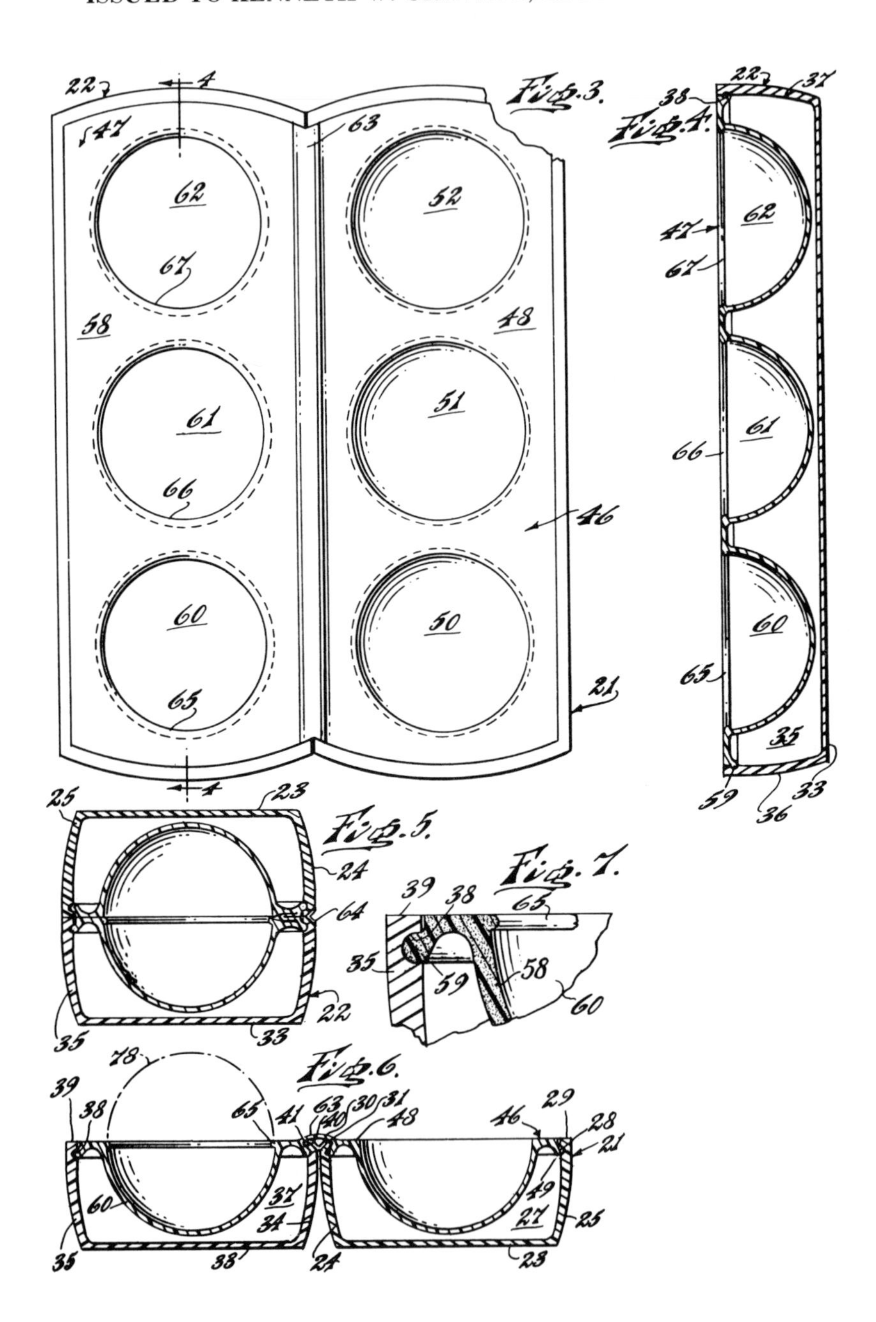

Even one's golf game can be improved by using solar energy. As most golfers know, golf balls travel farther when they are warmed. In Gravatt's device golf balls are placed in pockets (60), (61), and (62) and a top cover (21) is closed over them to form a sealed air space. A transparent outer cover allows sunlight to enter the chambers surrounding the balls to warm them. The entire device can be mounted on the outside of a golf bag for maximum exposure to the sun.

FIBER-OPTICAL SOLAR COLLECTOR

U.S. PATENT 3,780,722

ISSUED TO CHARLES J. SWET, 25 DECEMBER 1973.

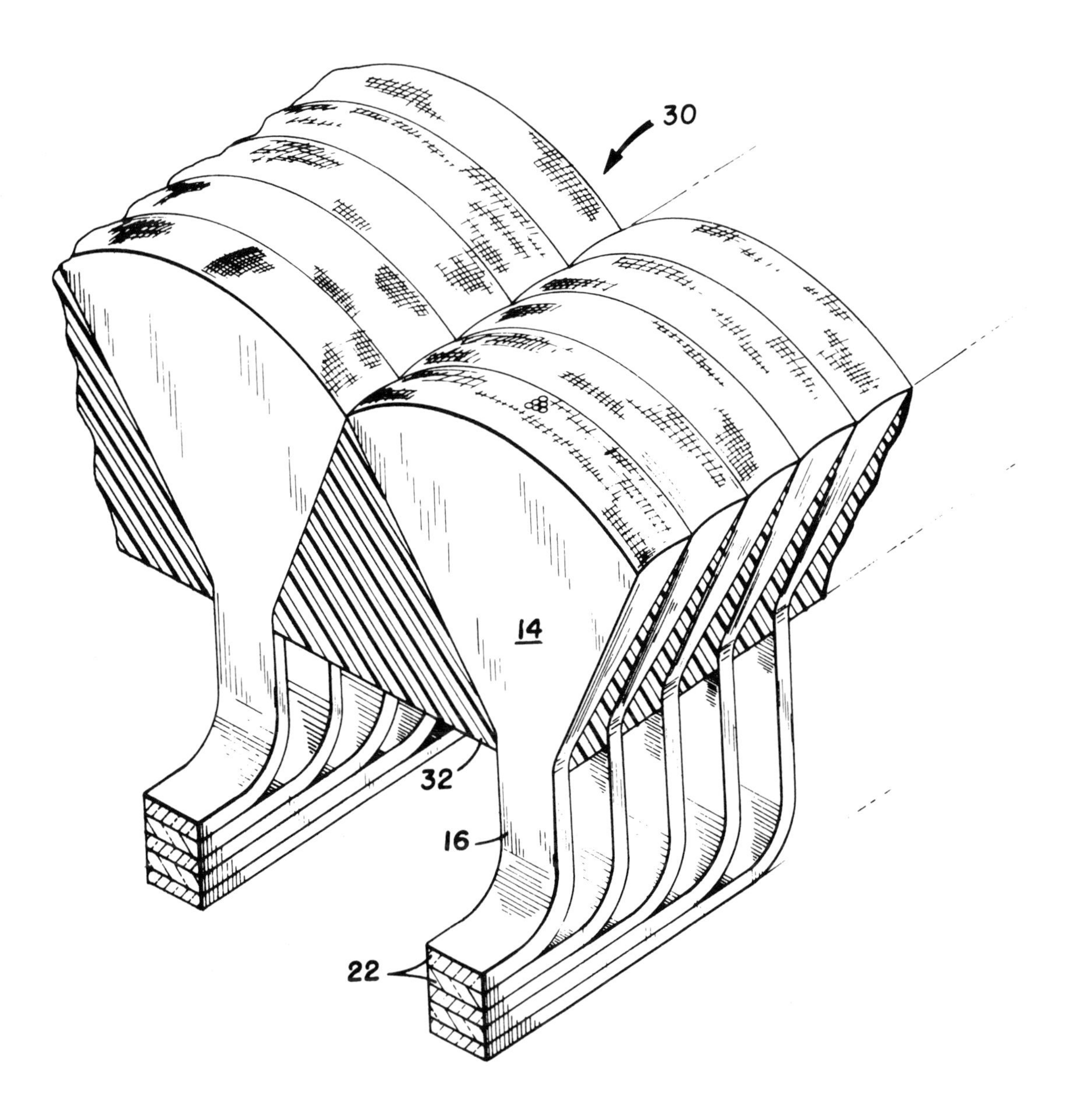

Not all sun-energy collectors require a fluid to transfer heat. By using solid instead of fluid sun-heat collectors, this device reduces both the operating costs and unit size, since it does not require pumping of liquid or air through internal flow channels as do most solar collectors. Multiple solid plastic fibers are bundled together at (14) to form a solar-absorbing surface. The separate bundles are joined at a lower point (22) to conduct collected heat to a storage mass or electric generator. The upper bundle surfaces are curved so that at least some fibers receive direct sunlight at all times of the day.

SOLAR HEATING SYSTEM

U.S. PATENT 3,815,574

ISSUED TO GEORGE R. GAYDOS, 11 JUNE 1974.

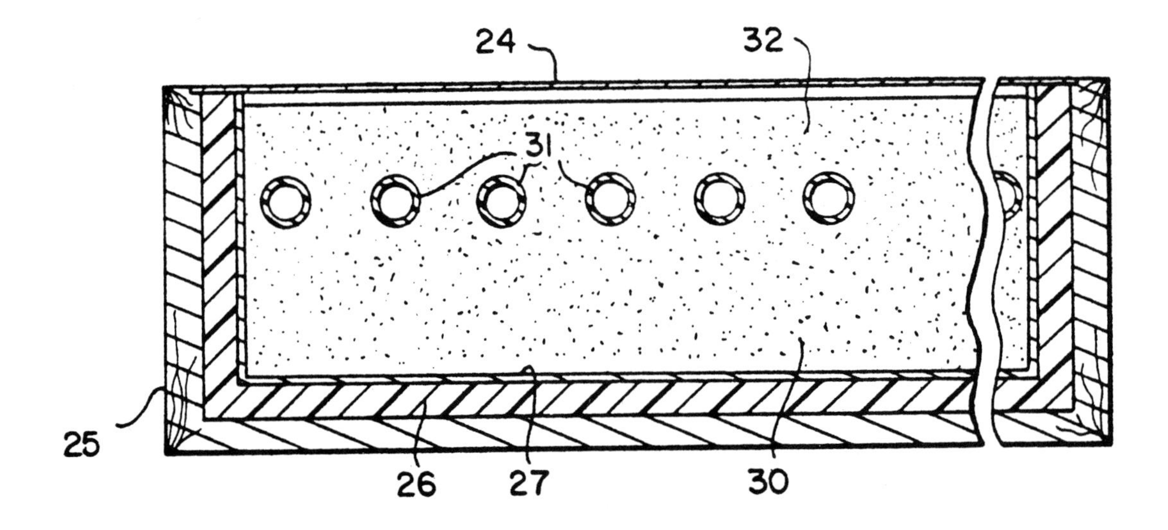

The use of cheap and available materials such as wood and sand simplifies the building of a solar collector. In this device, a heat-absorbing sand filler (32) surrounds water pipes (31). The outer frame (25) is made of wood. Insulation layer (26), liner (27), and glass cover (24) complete the structure. The invention shows how using materials that are easy to work with can benefit the mass production of solar collectors.

BUILDING STRUCTURES

U.S. PATENT 3,894,369

ISSUED TO ROBERT F. SCHMITT ET AL., 15 JULY 1975.

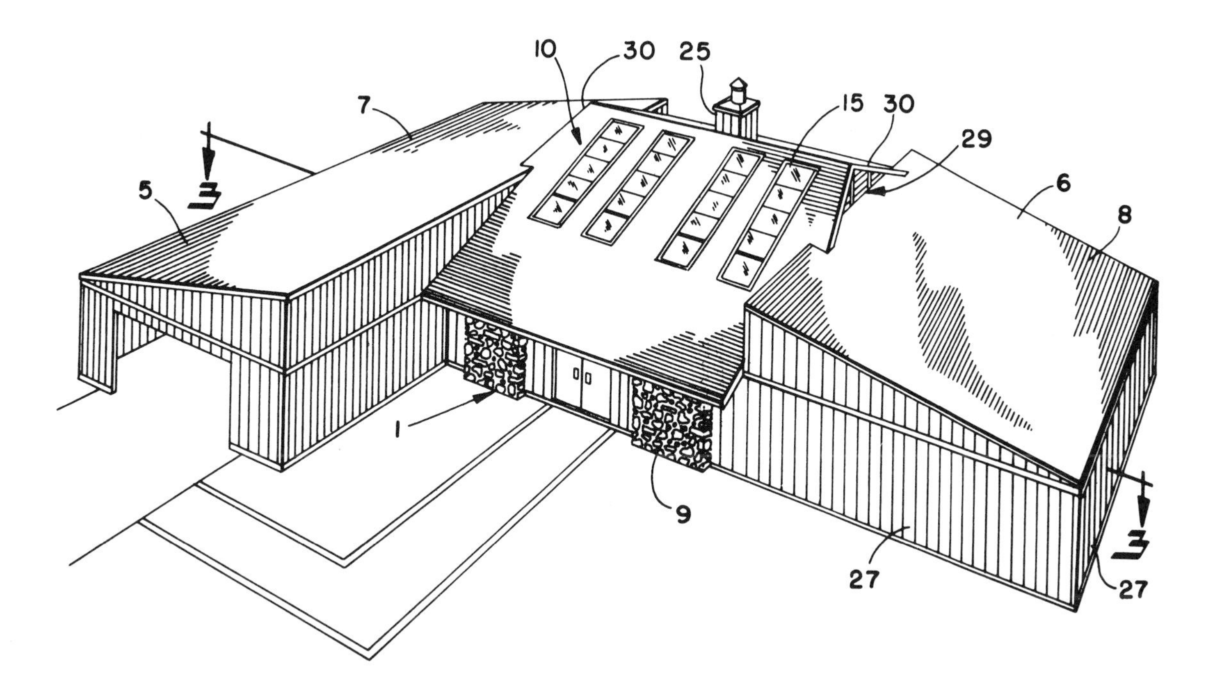

By placing windows on a slanted roof, a home becomes a more efficient solar collector. During sunny winter days, the central room (10) is heated by solar rays passing through windows (15); this heat is then transferred to the rest of the home. On cloudy days, room (10) is isolated from the rest of the house by simply closing doors. In summer, the solar heating effect is reduced by opening outside doors and windows to create a breezeway effect.

SOLAR PREHEAT CHAMBER FOR GRAIN DRYERS

U.S. PATENT 3,919,784

ISSUED TO MARTIN H. TONN, 18 NOVEMBER 1975.

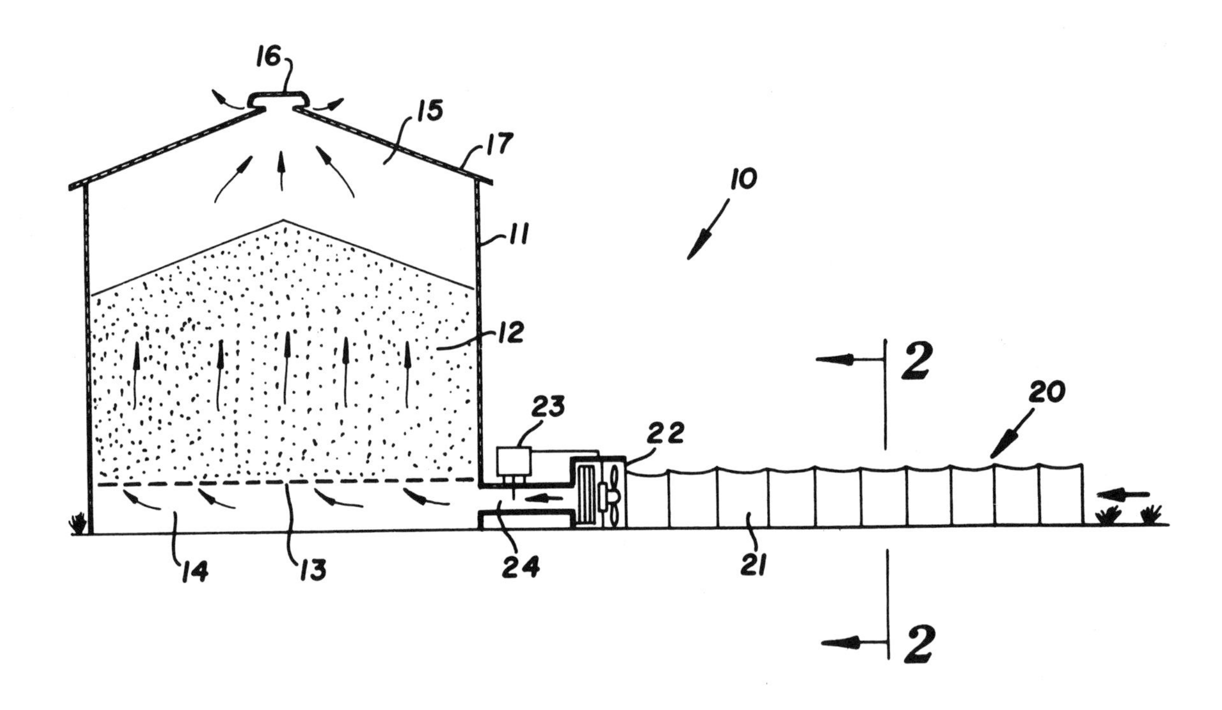

In addition to providing warmth and nutrition for crops, solar energy can also be used to dry them after harvesting. Grain (12) to be dried is stored in bin (11). Solar energy is captured by a long clear plastic cover (20) in area (21) to preheat the drying air. Fan (22) pulls the heated air through pipe (24). This heated air rises through perforated plate (13) and into the grain to dry it. Any excess heat exits at vent (16). This system eliminates the need for a costly gas burner to preheat the drying air.

METHOD FOR SUPPRESSING THE FORMATION OF ICE IN BODIES OF WATER

U.S. PATENT 3,932,997

ISSUED TO JOHN S. BEST, 20 JANUARY 1976.

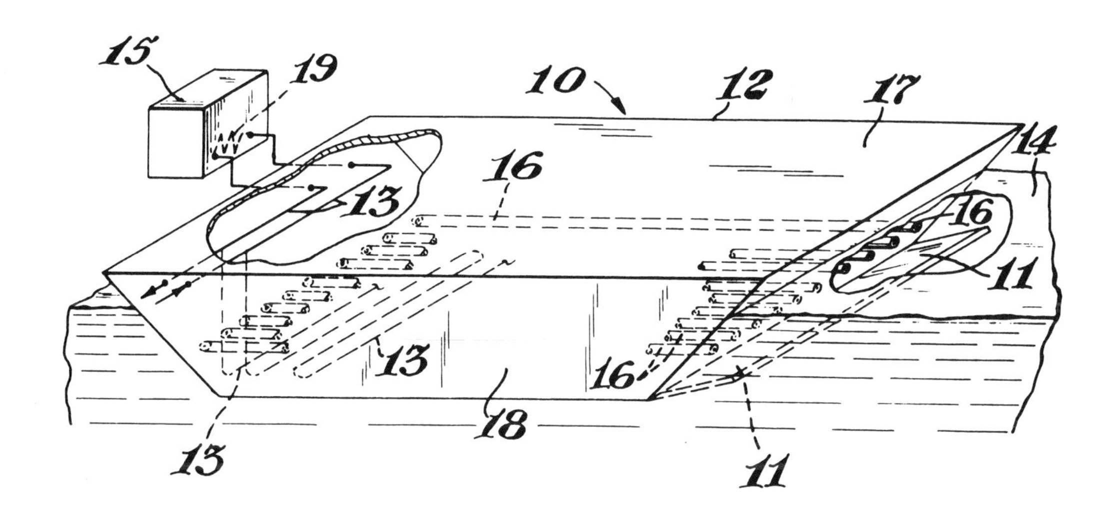

Ice formation on rivers can stop or hinder shipping traffic. Best's invention uses solar energy to help prevent river water from freezing. The system employs a movable river or canal barge containing an antifreeze liquid (18). Flow control members (11) allow near-freezing river water to flow into tubes (16). These tubes are surrounded by the antifreeze liquid, which helps to warm the river water and prevent it from freezing. Solar energy is used to raise the temperature of the antifreeze liquid (18).

SOLAR ELECTRICAL-GENERATING SYSTEM

U.S. PATENT 3,965,683

ISSUED TO SYDNEY DIX, 29 JUNE 1976.

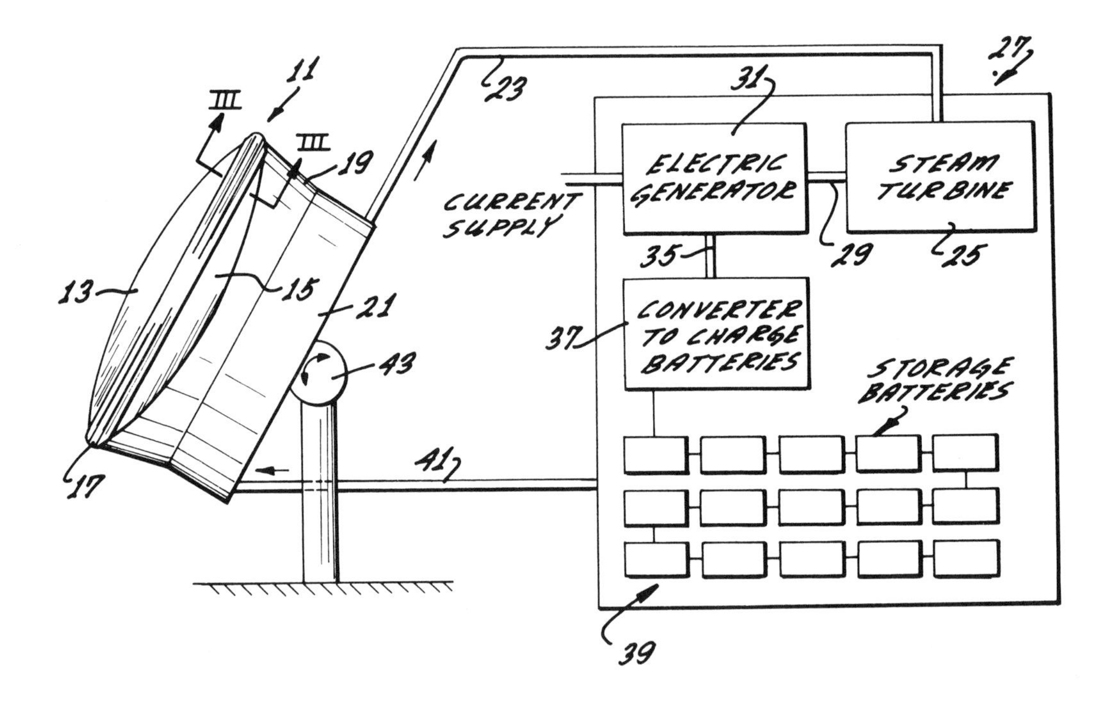

One of the main concerns in using solar energy is finding a way to store it for later use. Dix's invention uses batteries to accomplish this. Rotatable lens (13) allows sunlight to heat fluid (21). The steam generated flows via line (23) to steam turbine (25), which drives electric generator (31). The resulting electricity passes through converter (37) and then to batteries (39) for storage. One advantage of this system is its relatively small size as compared to the large solar collector panels required for many other designs.

HEATING AND COOLING SYSTEM

U.S. PATENT 3,965,972

ISSUED TO ROSS K. PETERSEN, 29 JUNE 1976.

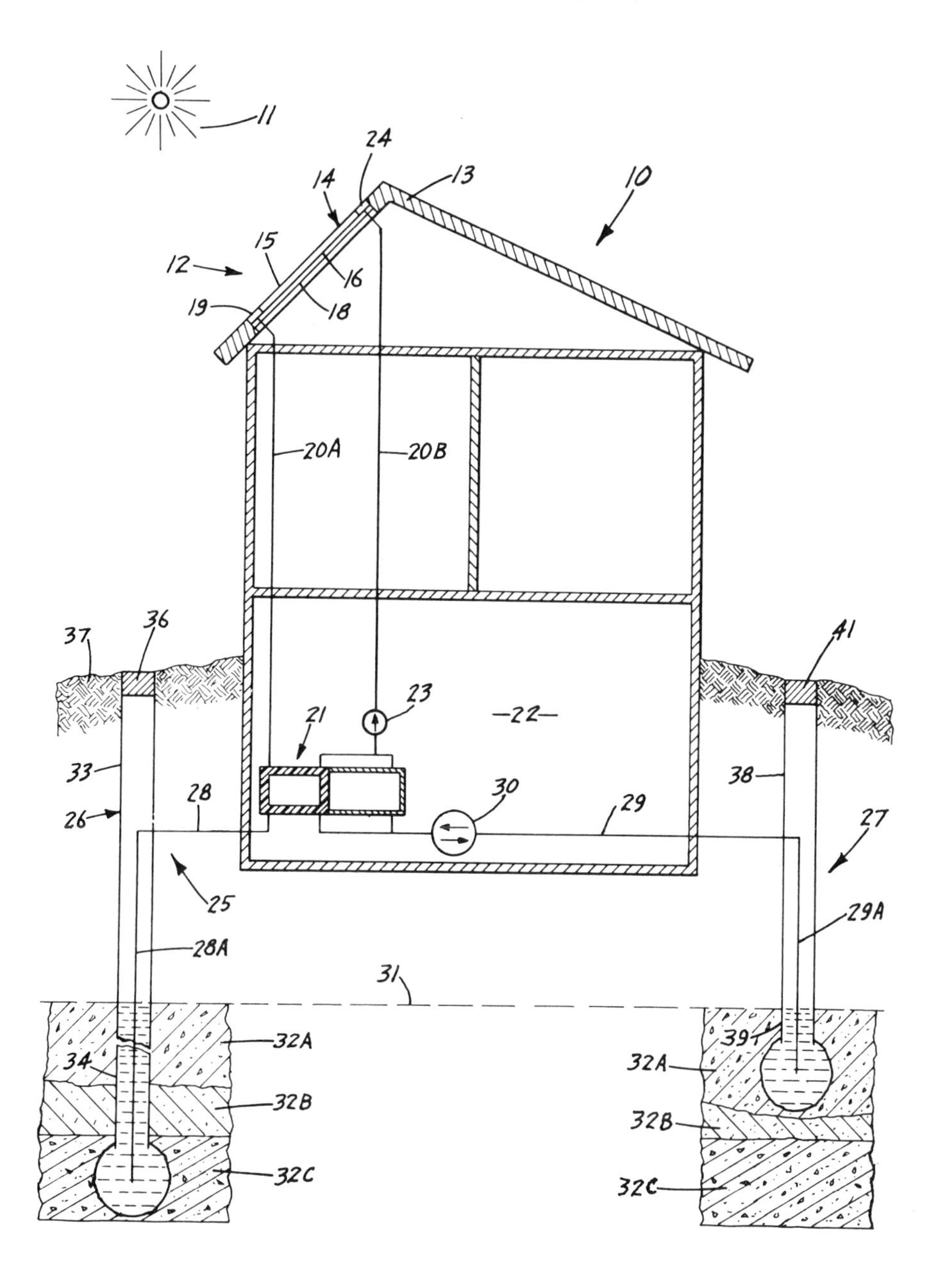

Existing water wells can be used for storage of solar energy, thus eliminating the need to build costly storage tanks beneath the earth. In Petersen's invention, two wells are used—one for storage of solar-heated fluid and the other for storage of cold fluids. In well (26), hot fluid that has been previously heated by the roof solar collector is stored in a deep underground location at levels (32) for later use. Cold-water well (27) supplies needed cold water that, if desired, can be preheated in a heat exchanger (21) before use.

41

SOLAR HEAT-ABSORBING TUBING

U.S. PATENT 3,968,786

ISSUED TO DAVID H. SPIELBERG, 13 JULY 1976.

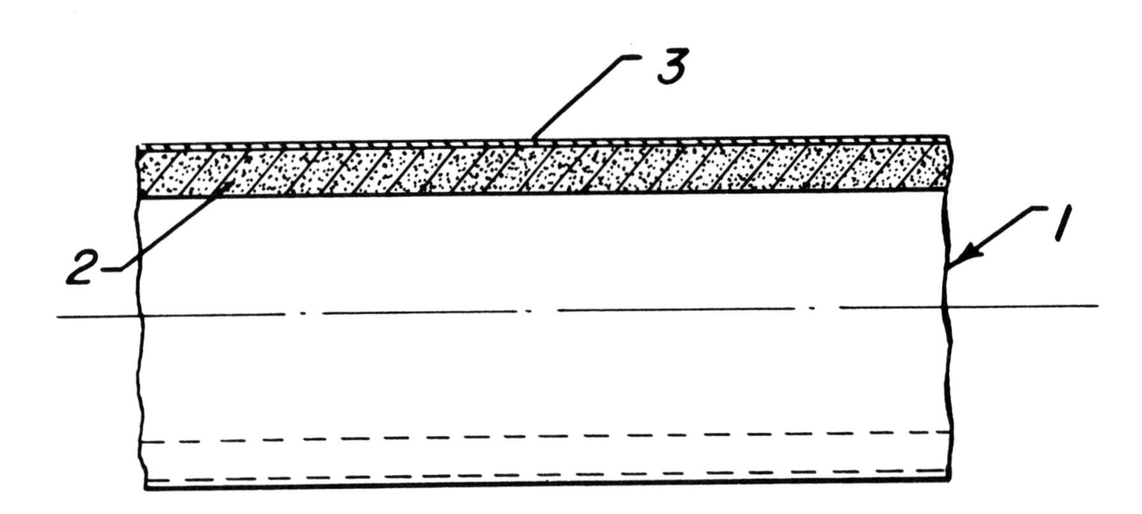

Plastic tubing is often used in solar systems to absorb and transfer heat. Spielberg's invention consists of manufactured plastic tubing that contains absorbing carbon particles (2) within the plastic walls (3). These carbon particles, being dark in color, will absorb solar energy more efficiently and will transfer this added heat into a fluid (1). The tubing reduces at-site installation costs, since it already contains the absorbing carbon particles.

COMBINATION SOLAR HEAT COLLECTOR AND AWNING

U.S. PATENT 3,973,553

ISSUED TO JOSEPH A. LANCIAULT, 10 AUGUST 1976.

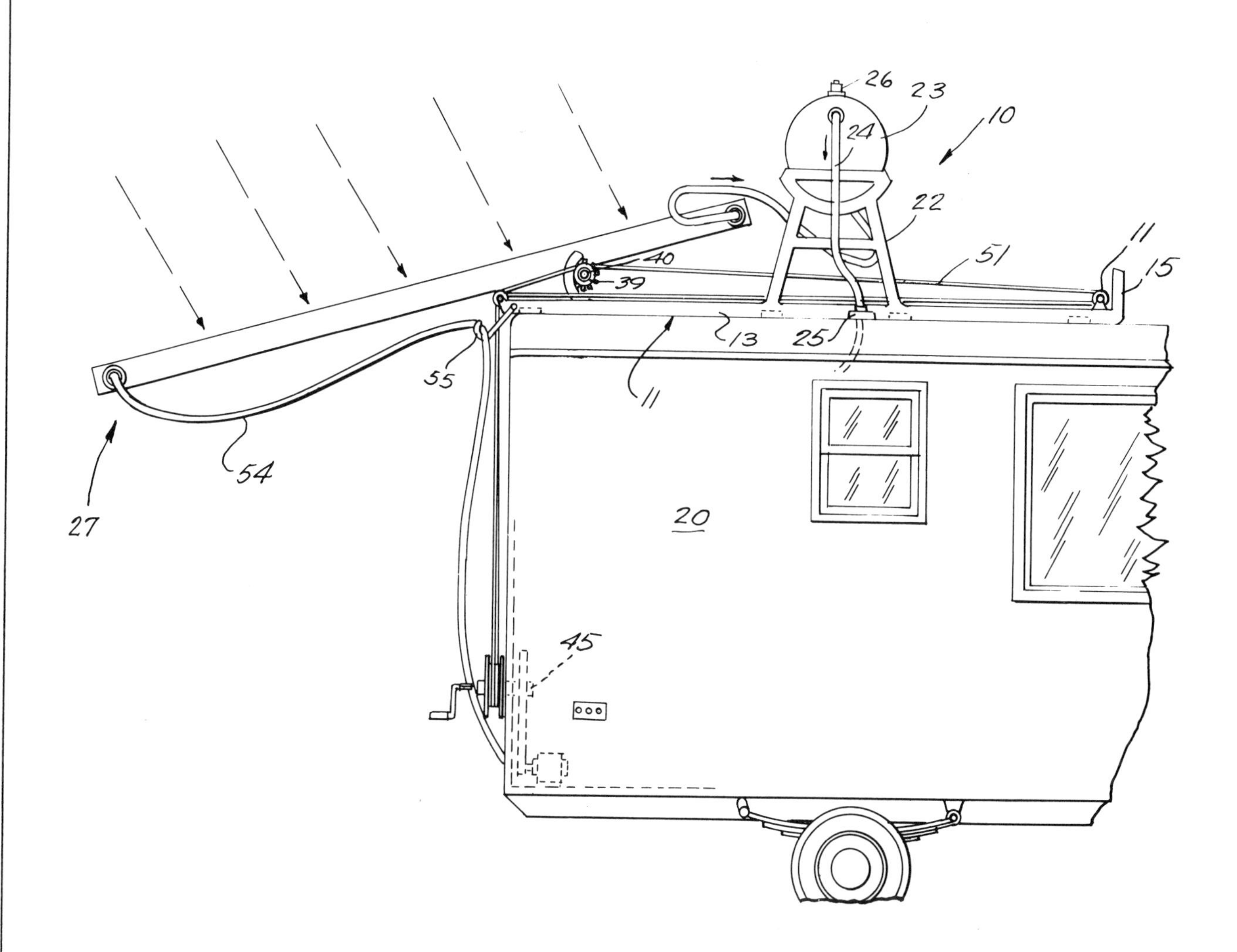

Some inventors have taken the concept of solar-heated homes and extended it to campers. Such a system could eliminate the need for costly and somewhat dangerous bottled gas heaters to provide heat and hot water. The collector is attached to the back of the camper and is angled for optimal solar energy pick-up by means of the crank handle and upper cable (51). Cold water, supplied to the collector via line (54), is heated in passages in the collector and is pumped to tank (23) for use as required in a domestic hot water system or in heat radiators. For travel purposes, the collector is lowered to the vertical position. In its extended position, the solar collector also acts as a shade awning for the camper.

43

SOLAR-ENERGIZED STEAM GENERATOR SYSTEM

U.S. PATENT 3,993,041

ISSUED TO WILLIAM F. DIGGS, 23 NOVEMBER 1976.

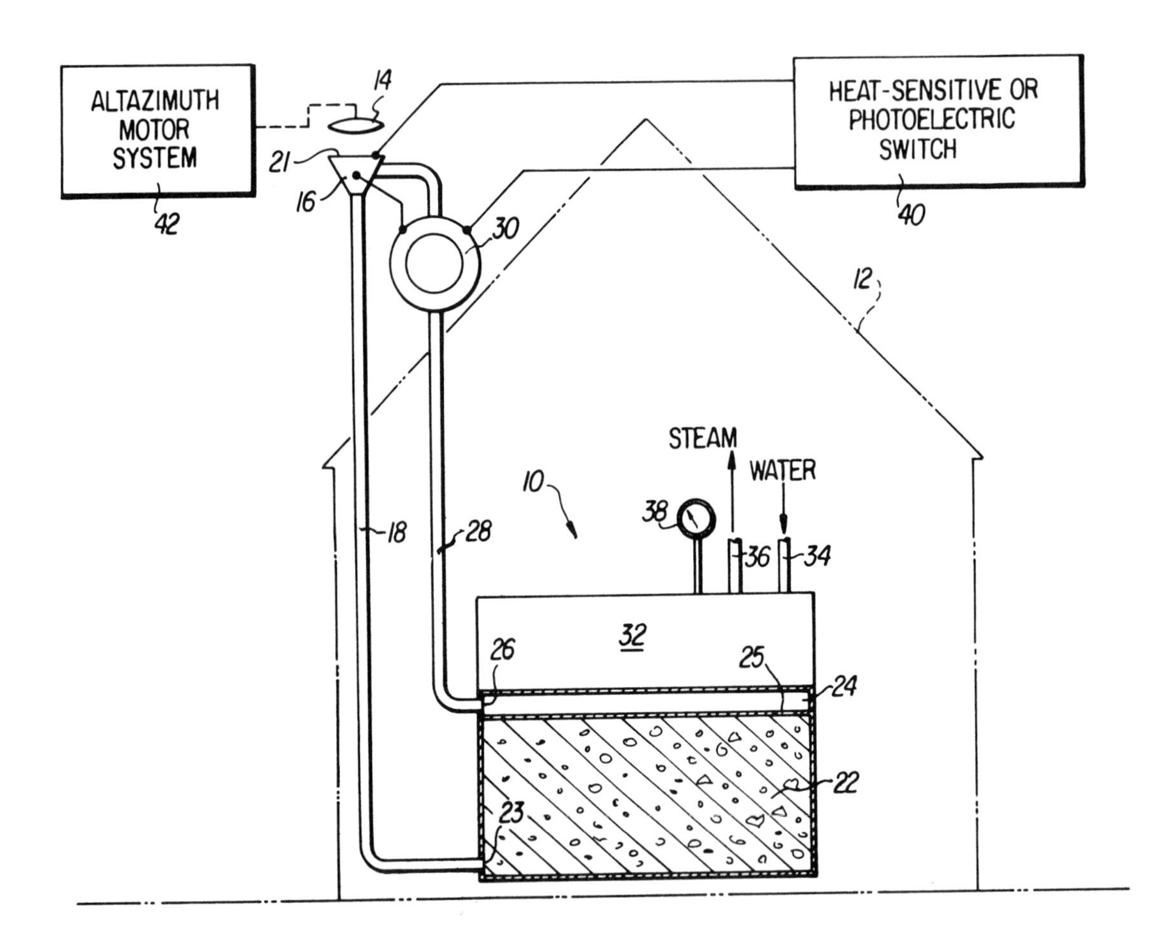

Diggs shows some of the many variables that can be used to control a solar collector. In his system, air driven by fan (30) is passed into a solar collection area (16) and, once heated, flows down through pipe (18) into tank (32), where it is used to generate steam. The fan (30) is operated when sensor (40) detects either a high temperature or the presence of sunlight in the collector (16). Also provided is a motor (42) that moves lens (14) to track the sun's position.

SOLAR-HEATED AND COOLED MODULAR BUILDING

U.S. PATENT 4,000,850

ISSUED TO RICHARD E. DIGGS, 4 JANUARY 1977.

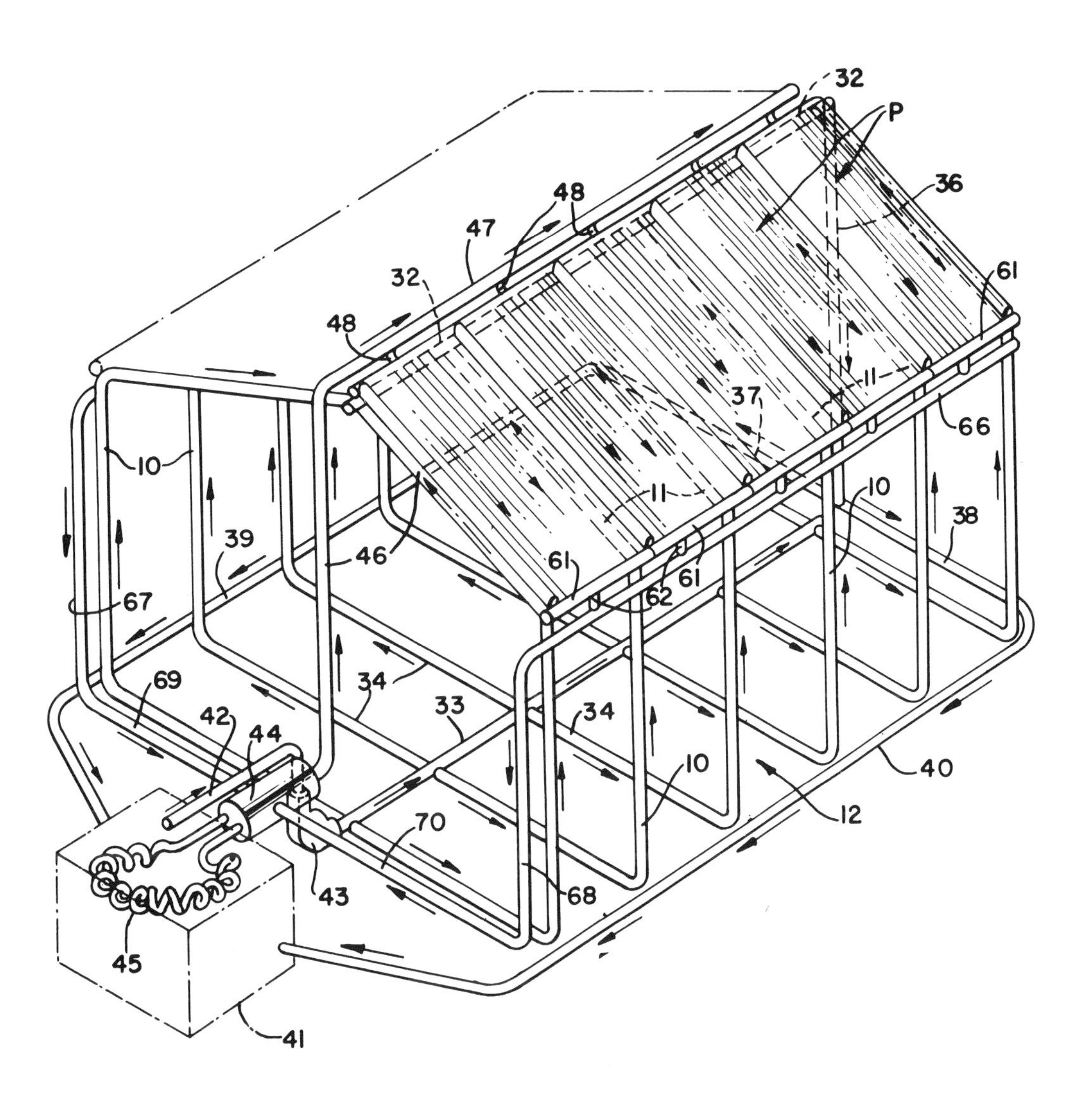

The Diggs invention shows the concept of using solar energy in prefabricated solar homes. The prefabricated walls contain fluid-distribution tubes (10). The building roof includes solar collector panels and distribution pipes (47) and (61). Pump (43) distributes solar-heated fluid to the wall tubes (10) to evenly heat the building via radiant interior panels. This idea would reduce on-site construction costs incurred when conventional solar collectors are installed. As an added benefit, the multiple fluid lines will also render the building essentially fireproof.

SOLAR-ENERGY CONVERTER WITH WASTE-HEAT ENGINE

U.S. PATENT 4,002,031

ISSUED TO RONALD L. BELL, 11 JANUARY 1977.

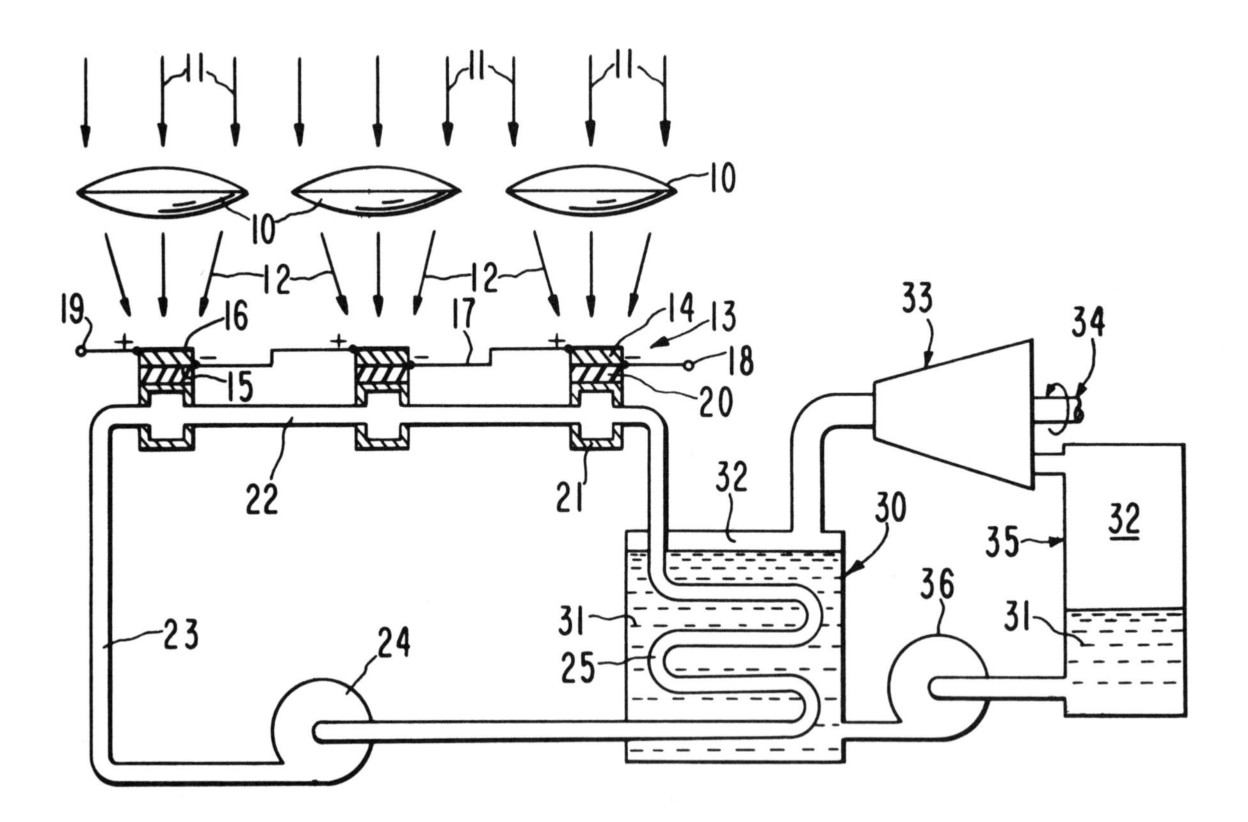

Many solar systems operate using energy collected in photovoltaic cells or in a solar-heated fluid. Bell's invention combines these two systems to accomplish a more efficient conversion of solar energy into usable energy. The sun's rays are focused through lenses (10) and directed onto photovoltaic cells (13), which convert sunlight directly into electric current. This current flows through lines (17), (18), and (19). In addition, excess heat from solar cells (13) passes to heat exchangers (21), where it heats fluid in pipe (22). This heated fluid is then pumped into tank (31) and is used to indirectly power an engine (33).

SUPPORT STRUCTURE FOR SOLAR-ENERGY CONVERTER

U.S. PATENT 4,002,158

ISSUED TO REINHART RADEBOLD, 11 JANUARY 1977.

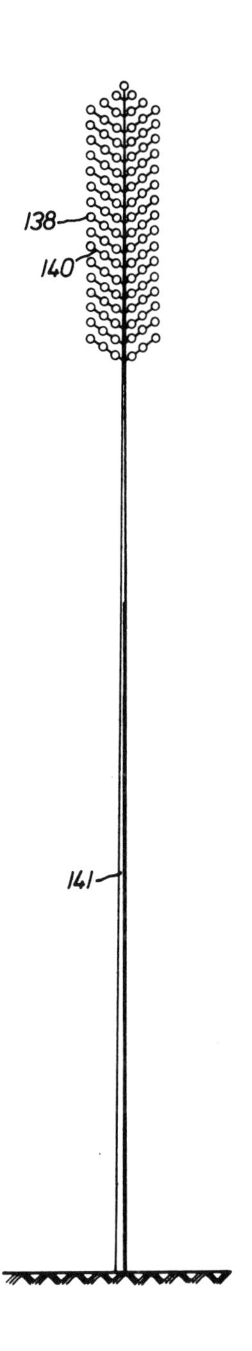

This tree-shaped solar collector could be used in locations where sunlight is blocked out at lower levels—such as in areas of densely crowded buildings. Pole (141) is of great height and supports stems (140), which have solar collector units (138) extending from them. Energy is collected in units (138) and transmitted via pole (141) to the ground for use.

VENETIAN BLIND FOR SOLAR HEATING

U.S. PATENT 4,002,159

ISSUED TO DOMENICK J. ANGILLETTA, 11 JANUARY 1977.

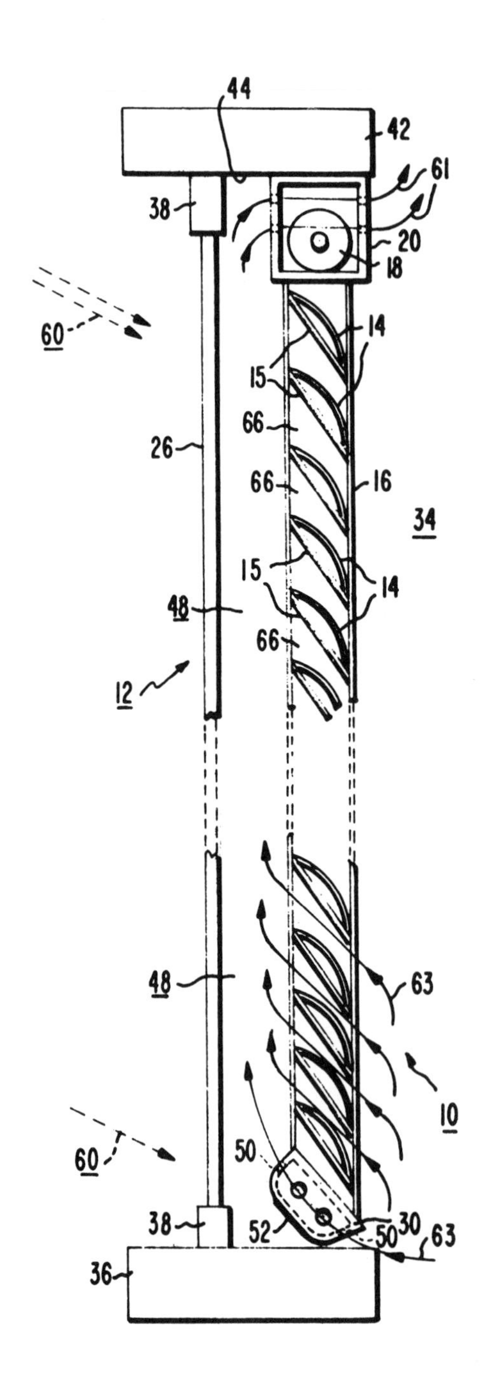

Ordinary venetian blinds can be easily modified into a solar collecting system. The device uses a blackened side (14) to absorb sun energy in winter and a white side (15) to reflect sun in the summer. Air flow to the room from space (48) is improved by drilling holes in rails (20) and (30), which permit lower air entry at (63) and upper air exit at (61).

DOUBLE-WALLED INFLATABLE STRUCTURES

U.S. PATENT 4,004,380

ISSUED TO JOHN P. KWAKE, 25 JANUARY 1977.

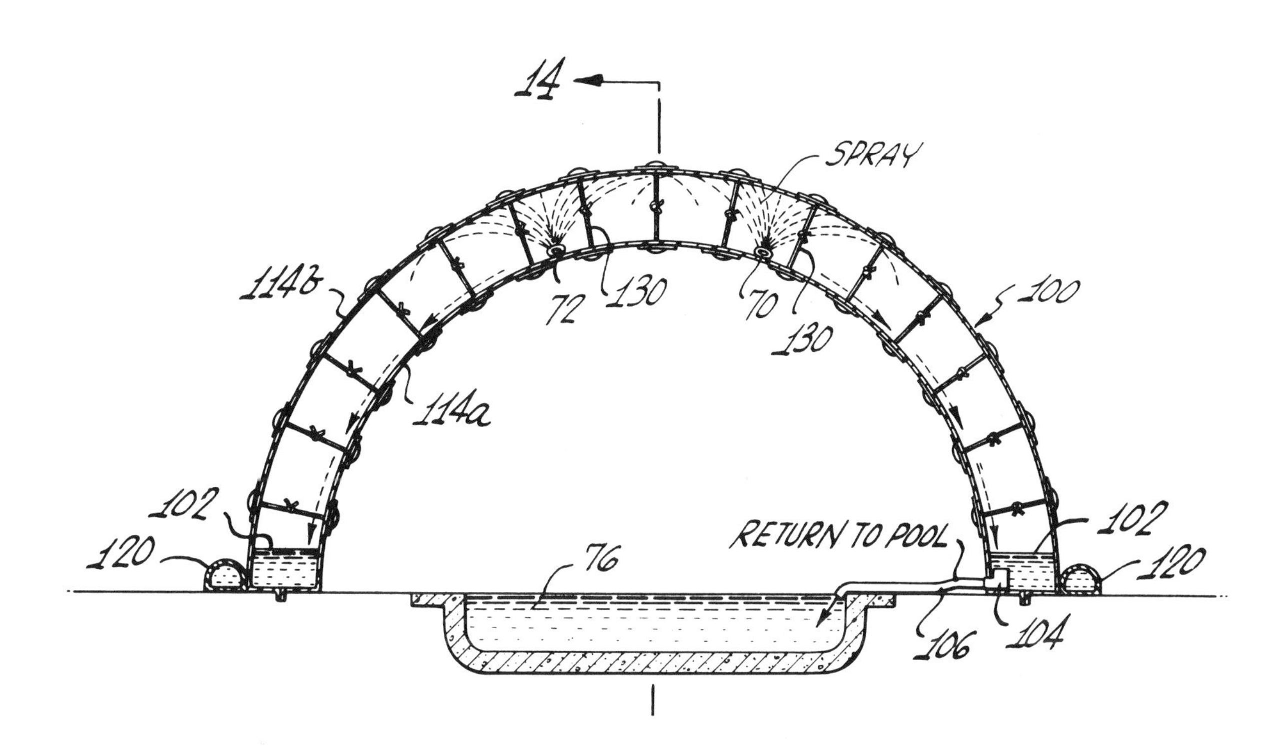

A temporary structure, such as a covered swimming pool, carport, or airplane hangar, may also be used as a solar collector. In this example, clear plastic walls (114a) and (114b) form a building and surround a pool of water (76). Sprays of water are delivered at (70) and (72) and trickle down along lower wall (114a) to pick up solar heat energy. The heated water then flows into chambers (102) for delivery to the pool. By using the sprays (70) and (72), the pool water is more quickly heated.

SYSTEM FOR CONVERTING AND STORING COLLECTED SOLAR ENERGY

U.S. PATENT 4,010,614

ISSUED TO DAVID M. ARTHUR, 8 MARCH 1977.

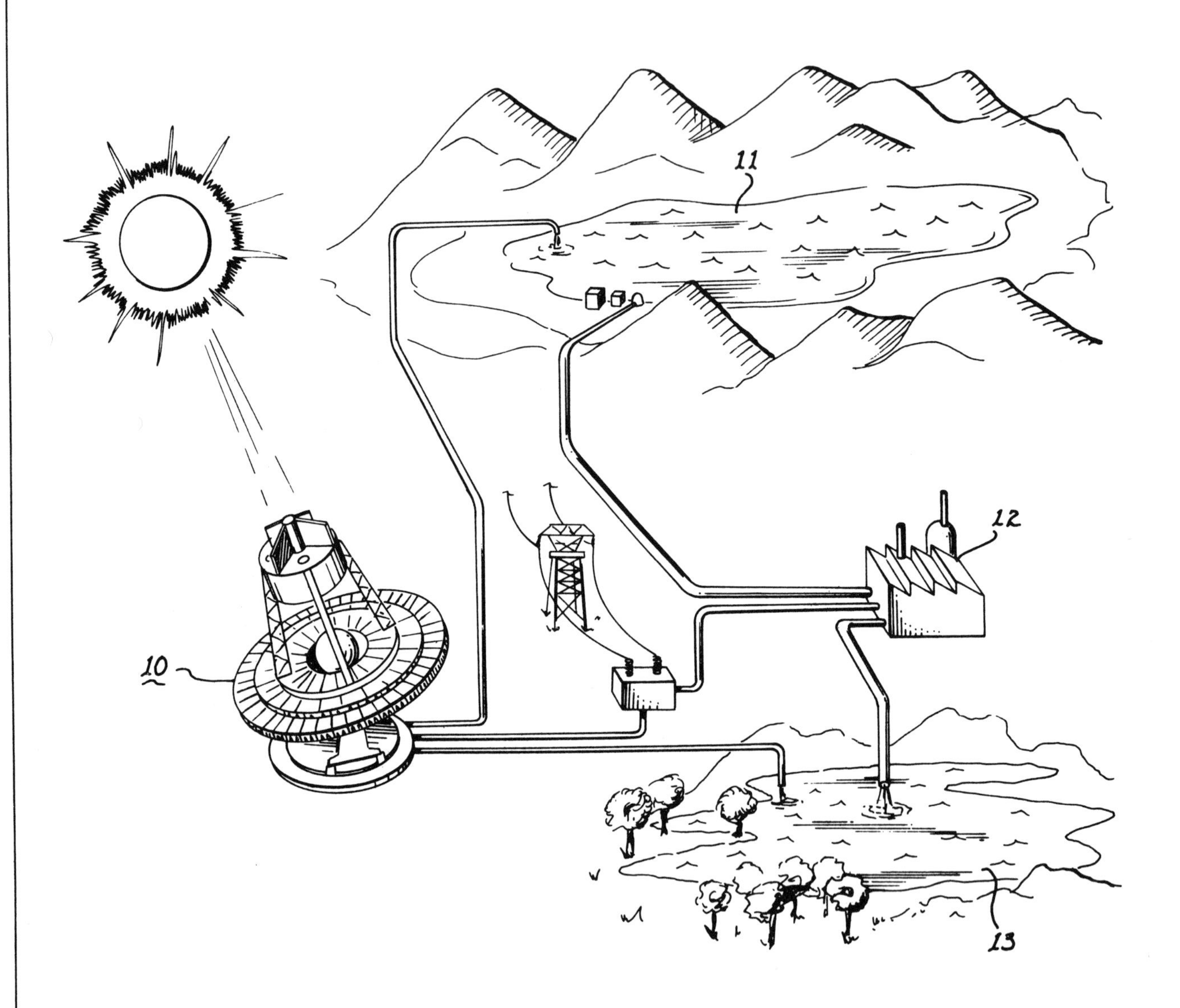

Future solar energy plants may be located near large natural or man-made ponds. In the Arthur system, sunlight is concentrated by collector (10) and used to heat or boil water. The heated fluid may then be stored in a pond (11) or it may be used immediately in a steam turbine to generate electricity for a town. If stored in pond (11), it may be used at a later time in a hydroelectric generator station (12) and the cooled fluid can then be discharged to a second pond (13).

SOLAR COLLECTOR

U.S. PATENT 4,011,855

ISSUED TO FRANK R. ESHELMAN, 15 MARCH 1977.

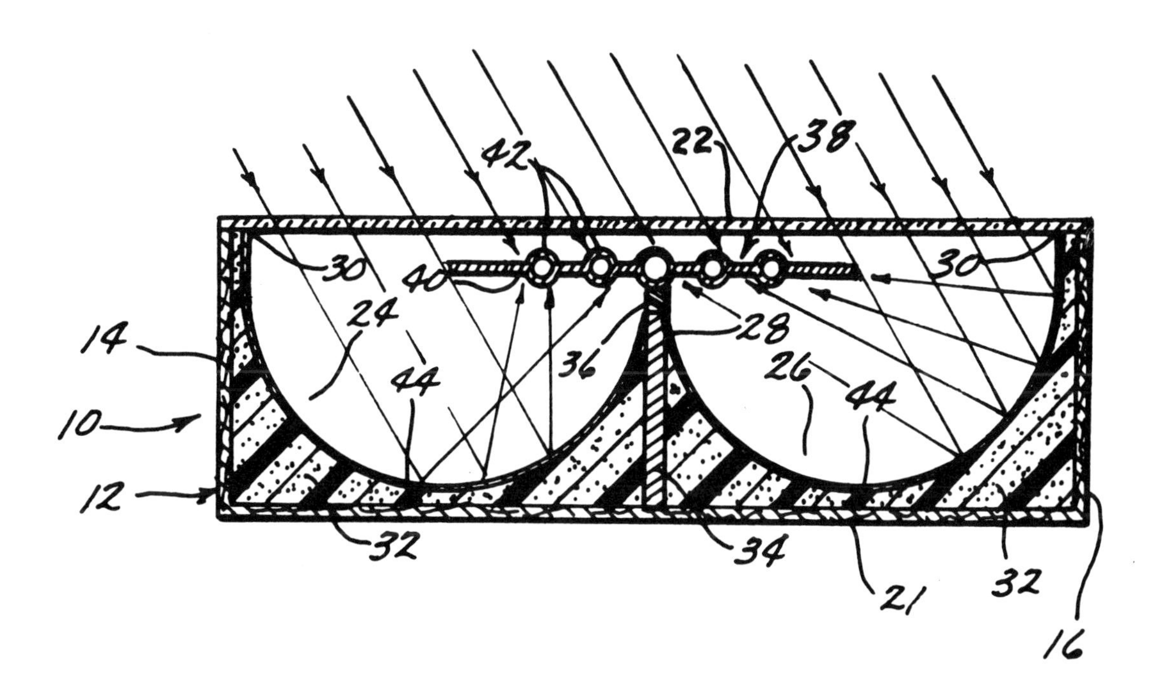

Even though it has no moving parts to track the sun, this collector gathers a large amount of the sun's energy due to the way it is shaped. Liquid-carrying tubes (42) are formed in a flat plate (38) and are placed on center stand (28). By using two curved reflectors (44), the absorber plate and tubes receive solar energy from both sides at most times of the day, thus increasing the heating effect. The lower part of the device is insulated by foam layers (32) and the upper part is sealed shut by glass plate (22).

TEMPERATURE CONTROL SYSTEM USING NATURALLY OCCURRING ENERGY SOURCES

U.S. PATENT 4,015,962

ISSUED TO LEO L. TOMPKINS, 5 APRIL 1977.

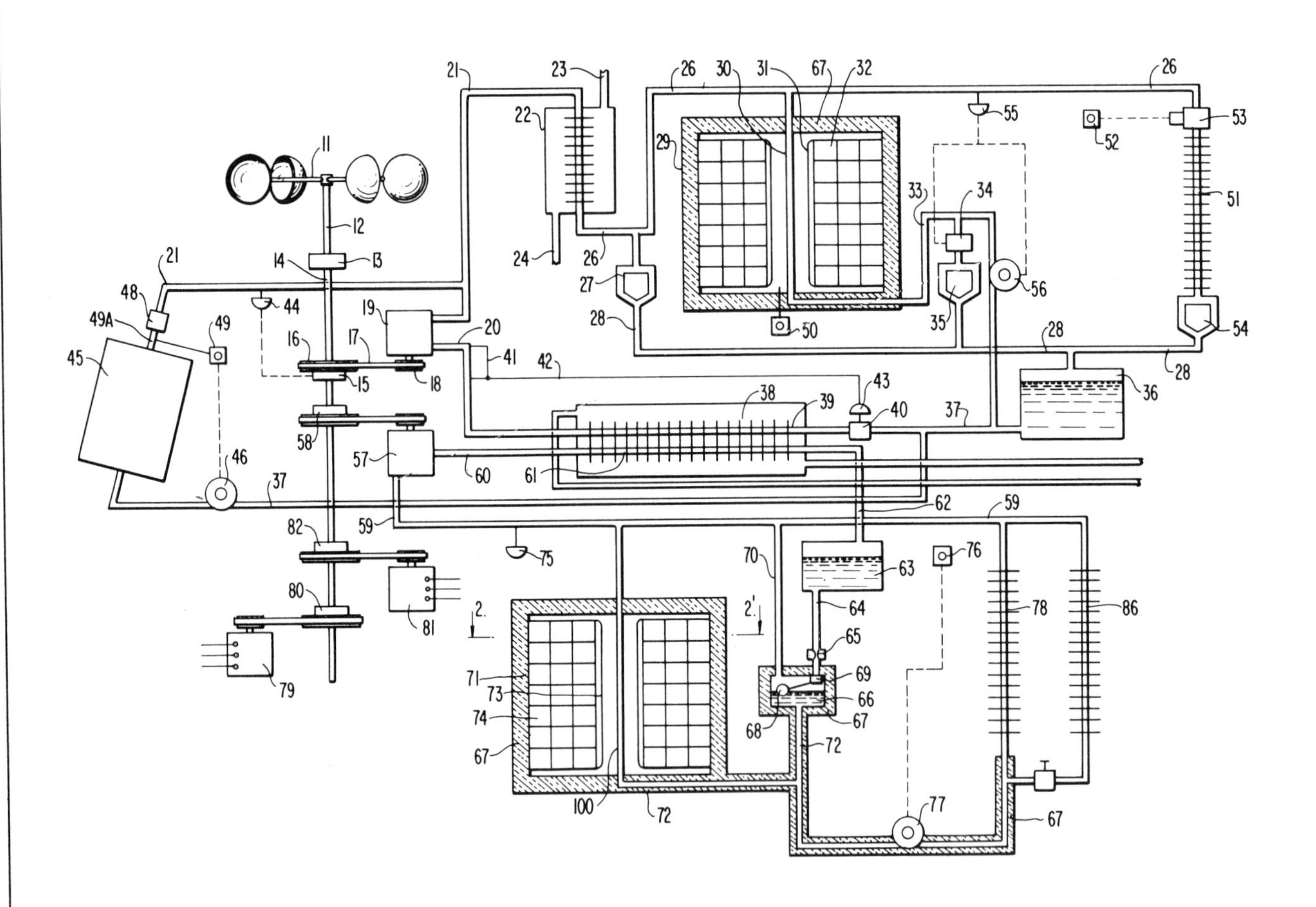

This invention combines two free energy sources, the sun and wind, into a single system by using a windmill to save money on pumping air through a solar collector. In operation, windmill (11) rotates to drive air fan (19) via drive belt (17). The compressed air from (19) then flows in line (21) to solar collector (45) (see left of drawing) where it is heated and flows to storage tank (29) or is used immediately in a radiator coil (51). The windmill (11) may also be used to rotate other fans on shaft (12) as part of a cooling system when needed.

SOLAR-POWERED PORTABLE CALCULATOR

U.S. PATENT 4,017,725

ISSUED TO STEPHEN A. ROEN, 12 APRIL 1977.

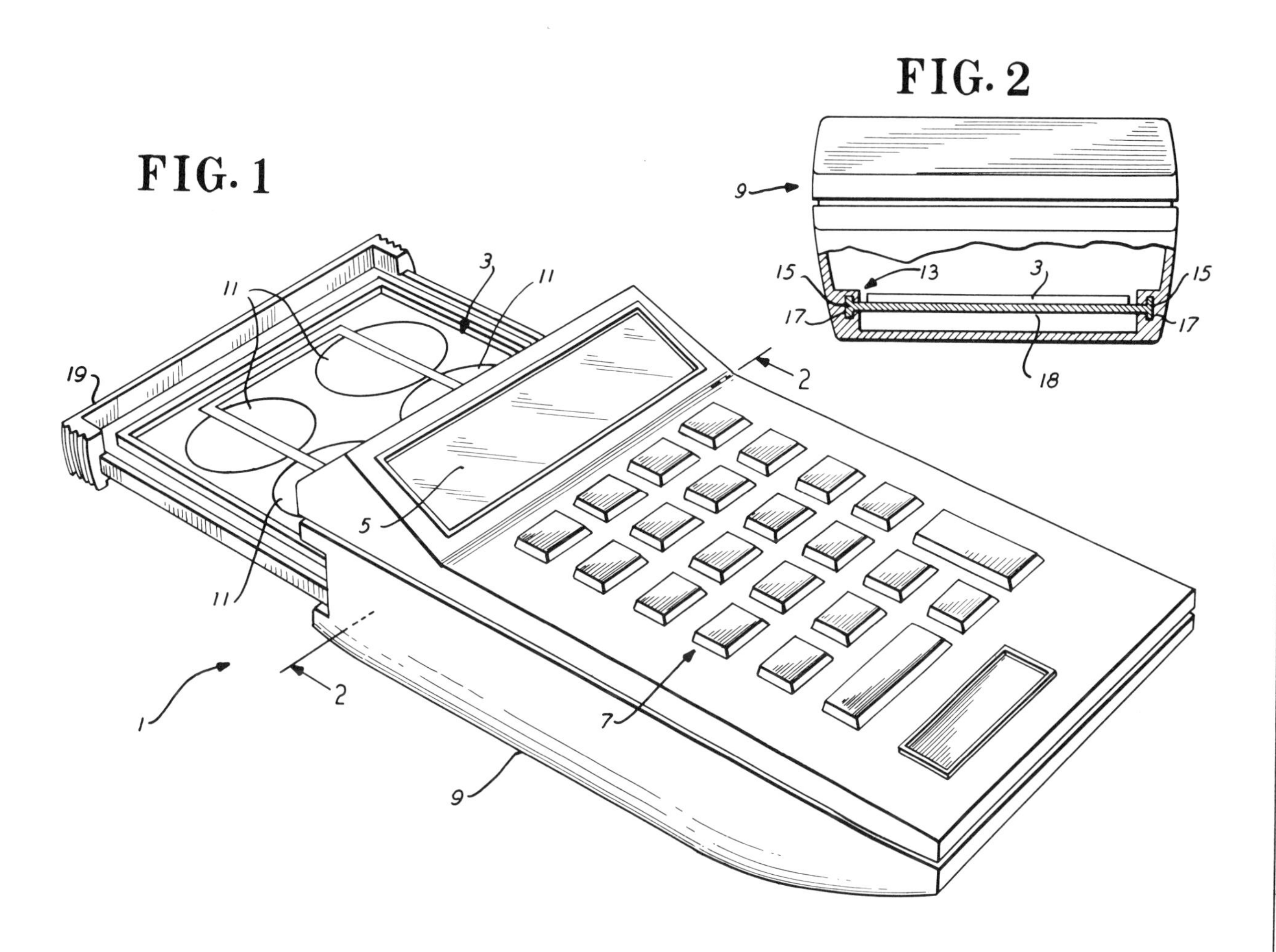

With Roen's solar rechargeable calculator there is no need to run to the store to buy batteries when the power is gone; simply set it in the sun. The calculator contains several solar cells (11) that are connected in series. When slide (19) is pulled out to expose the cells to sunlight, a current is generated and the calculator is recharged. Presently, many inventors are attempting to produce a more efficient solar (photovoltaic) cell. A breakthrough in this area would make devices such as Roen's very practical.

LIQUID-OPERATED SOLAR ENERGY COLLECTOR

U.S. PATENT 4,018,215

ISSUED TO YU K. PEI, 19 APRIL 1977.

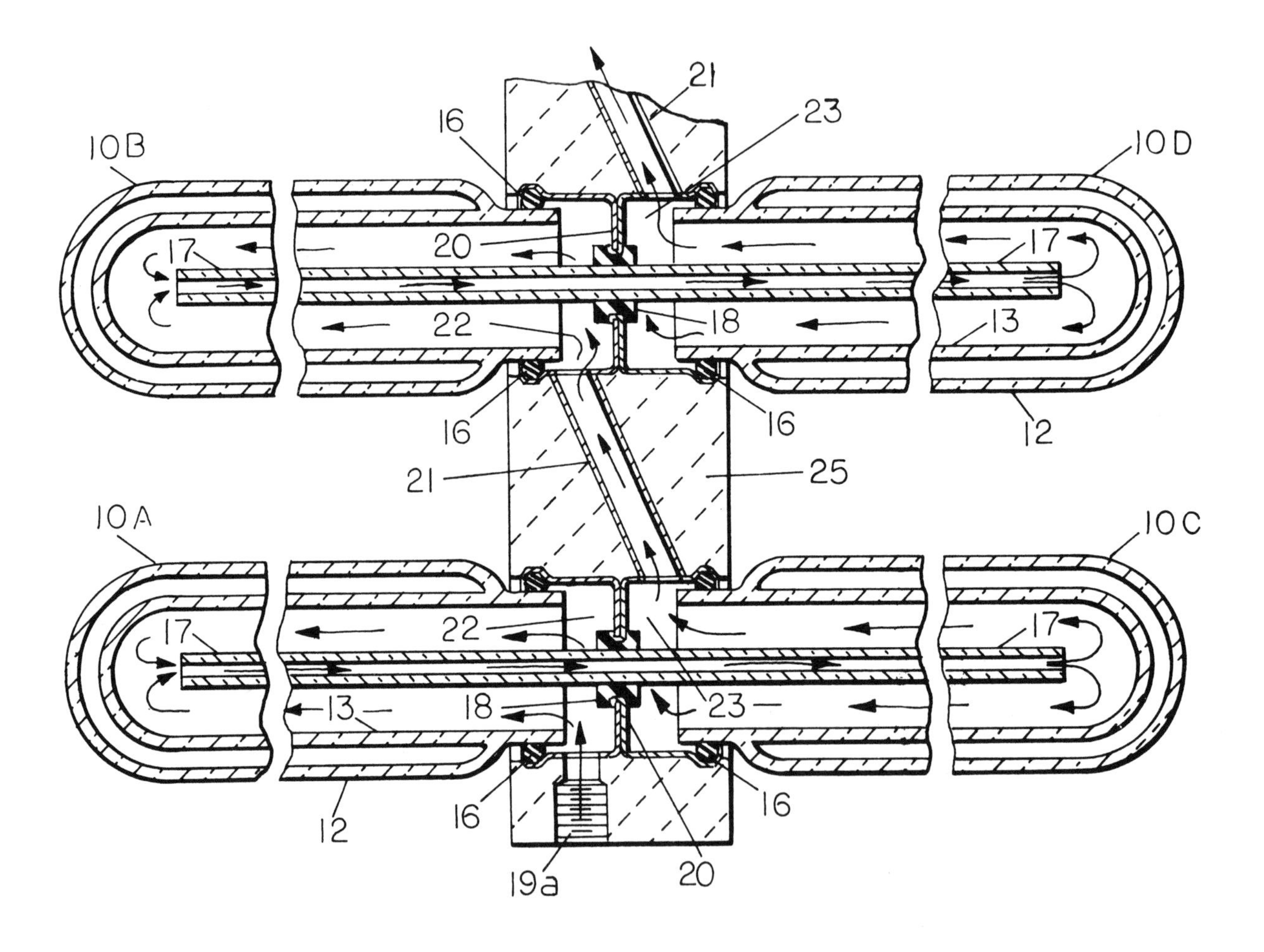

This system makes assembly and repair of a solar collector much easier. It uses glass pieces (12), (13), and (17) to provide a fluid flow path between (17) and (13) and an insulating air space between (12) and (13). The insulating air space helps keep more of the sun's heat energy inside the glass. Heated fluid flowing between separate collectors is insulated at (25). Since the glass pieces are simply held in place by seals (16) and (18), assembly and disassembly for repairs or cleaning can be done quickly.

ROOF UTILITY UNIT

U.S. PATENT 4,020,605

ISSUED TO STEVEN ZENOS, 3 MAY 1977.

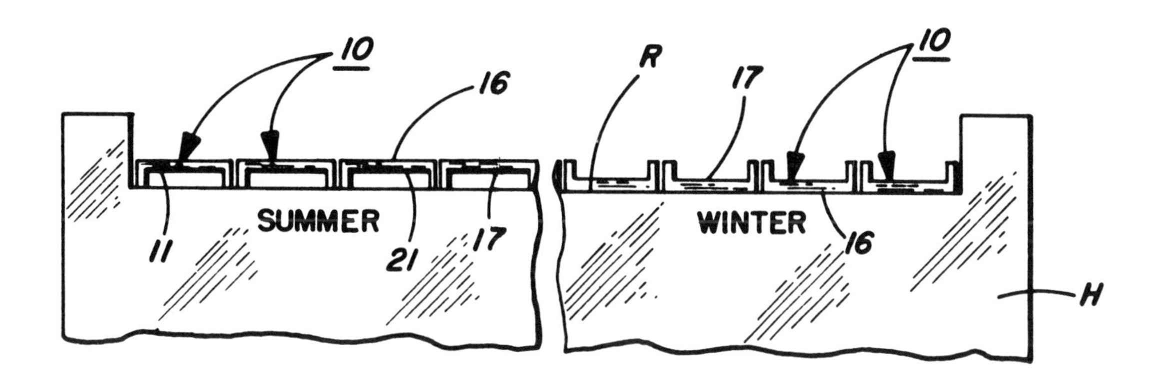

Heating and cooling bills can be reduced by changing the reflective properties of a roof to suit the season. Small wooden panels (approximately 2 ft. × 2 ft.) have a sun-reflecting white side (16) and a sun-absorbing dark side (17). In summer, reflecting side (16) faces up to repel solar rays and provide an insulating air space at (21). In winter, one person can easily reverse the small panels to have the sun-absorbing side (17) face up.

SOLAR-ENERGY SYSTEM

U.S. PATENT 4,020,826

ISSUED TO ROBERT A. MOLE, 3 MAY 1977.

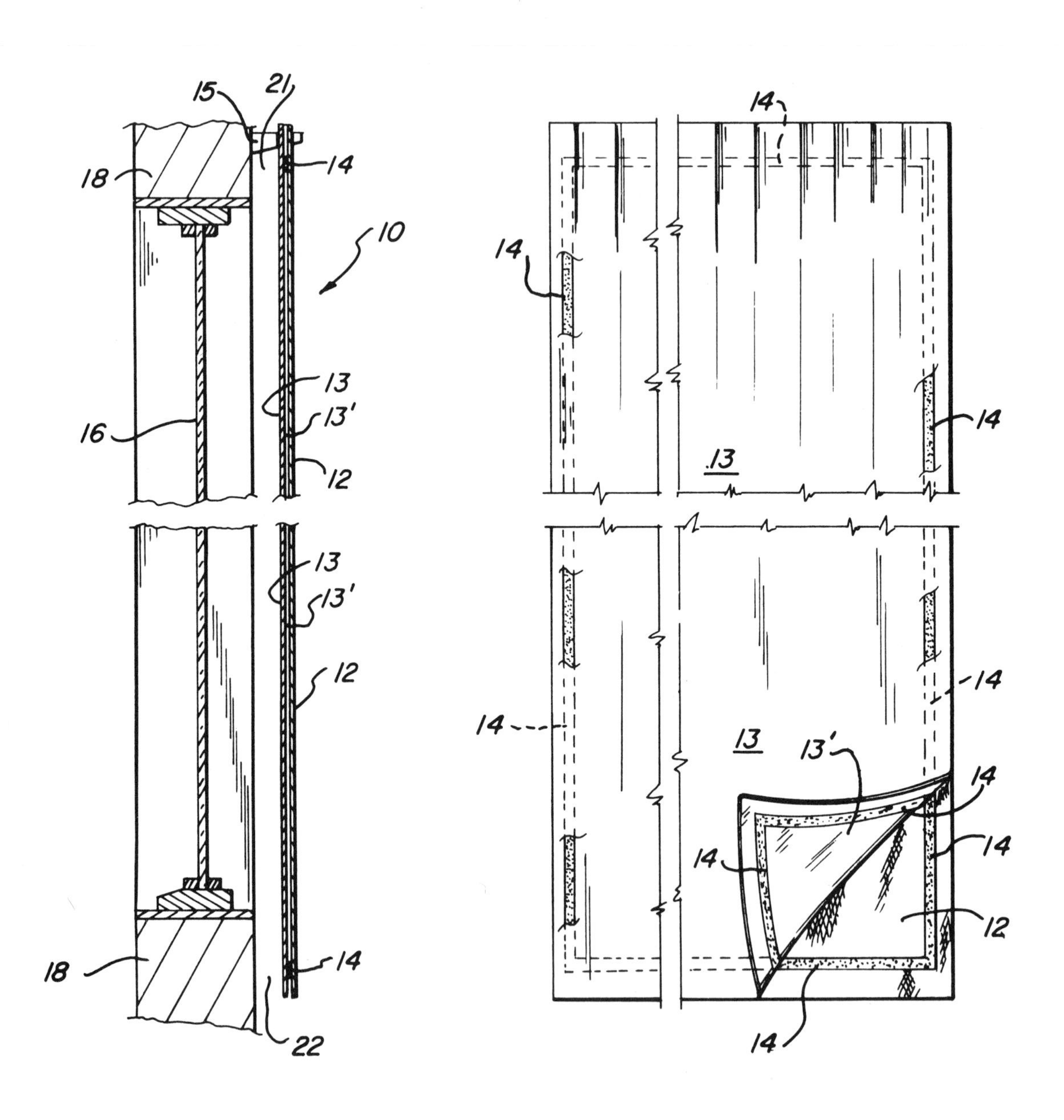

Window drapes can be designed to absorb the sun's rays when needed in winter and to reflect them in summer. Drape liner (13) is reversible to have a reflective side facing window (16) in summer and a dark heat-absorbing side facing the window in winter. Zippers (14) allow easy manual reversal of the liner. Note that the same drape (12) always faces the room interior. Extended curtain rods (15) provide air circulation paths (22) and (21) for removal of sun-heated air in winter months.

COMPOUND-LENS SOLAR-ENERGY SYSTEM

U.S. PATENT 4,022,186

ISSUED TO LEONARD L. NORTHRUP, JR., 10 MAY 1977.

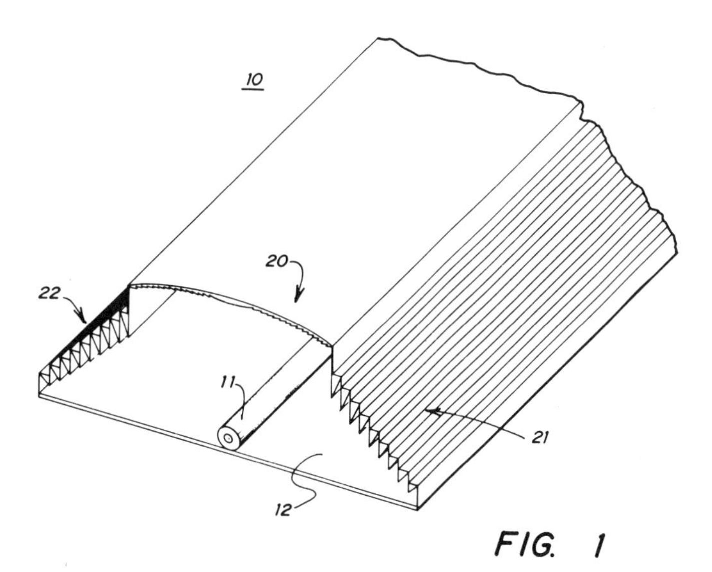

FIG. 1

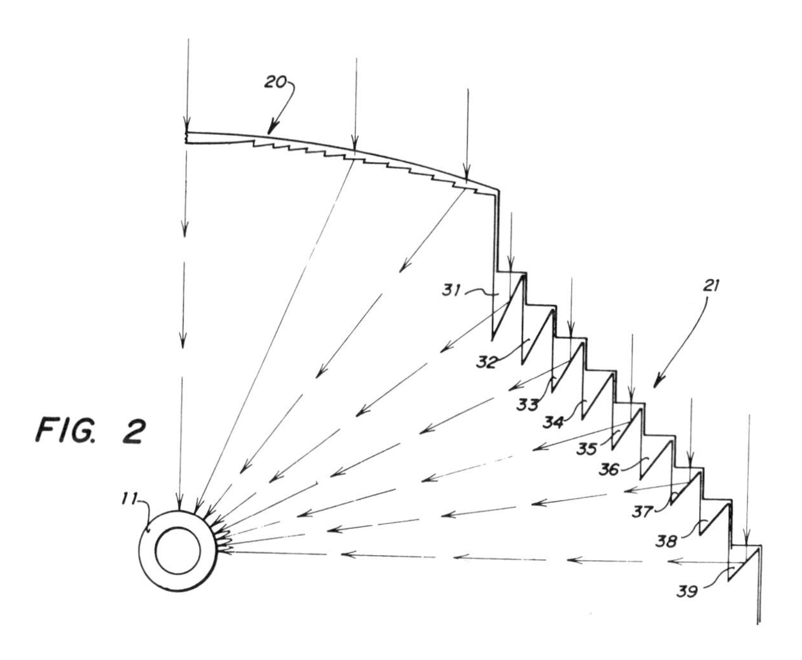

FIG. 2

In many solar-energy systems, lenses are used to concentrate the sun's rays. Northrup's system uses variously shaped lenses at (20), (21), and (22) to concentrate and direct sunlight toward a fluid-carrying tube (11). Since outer lenses (21) and (22) are shaped dfferently from central Fresnel lens (20), sunlight from a wider area can be captured and directed toward tube (11), thus heating the fluid within it more efficiently.

INFLATABLE BUILDING

U.S. PATENT 4,027,437

ISSUED TO JEFFREY N. MONSKY ET AL., 7 JUNE 1977.

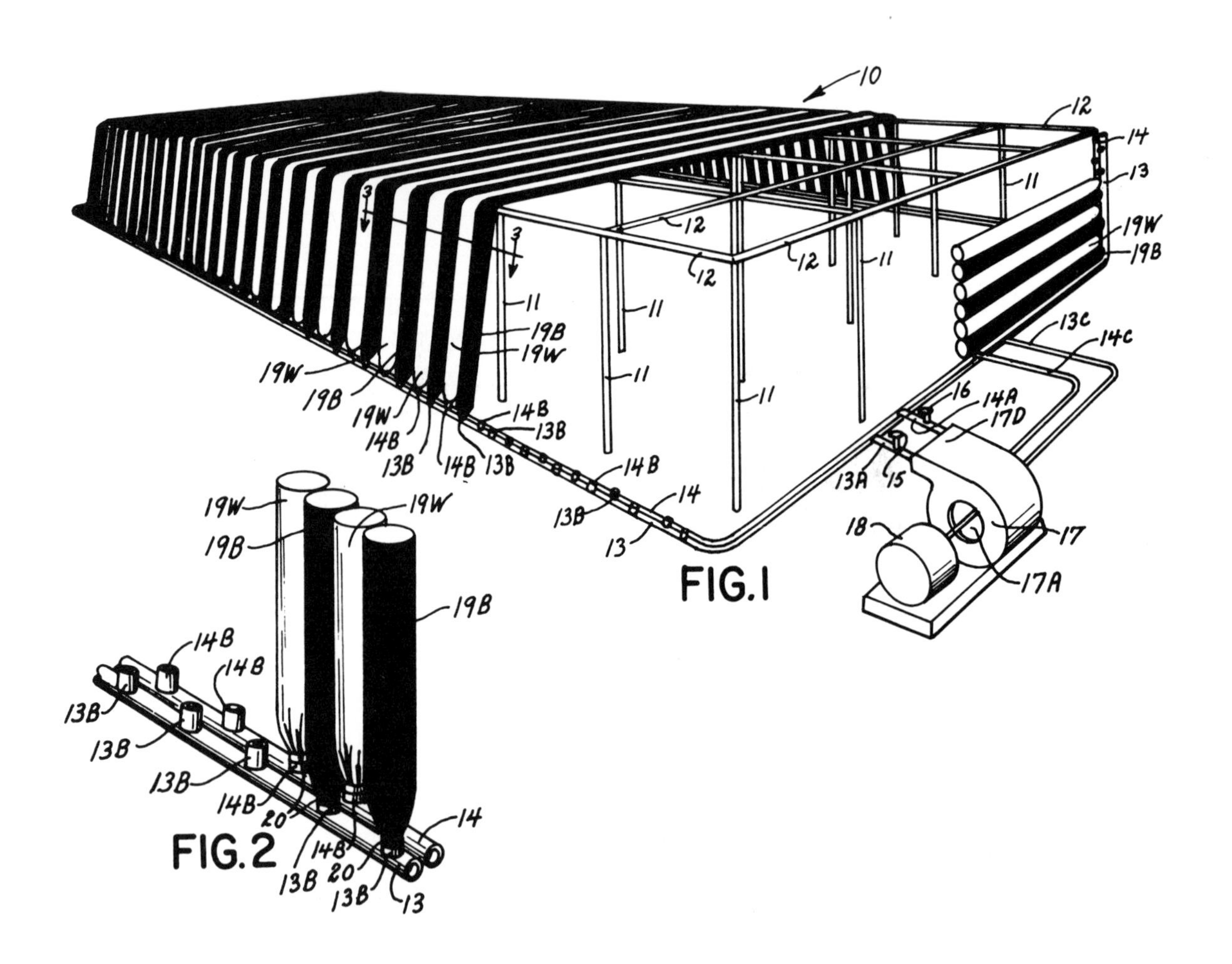

In this system a building is "blown up" in the same way we inflate a tire or balloon—by air pressure. Air supplied by fan (17) flows through tubes (13) and (14) to pump up the flexible outer walls. Monsky's design uses air valves (15) and (16) to change the solar properties of the structure. In summer, the white reflecting sections are inflated and the air is let out of the black sections. In winter, the black sun-absorbing walls are inflated and the white walls are deflated.

SOLAR-ENERGY COLLECTOR

U.S. PATENT 4,029,258

ISSUED TO GLENN F. GROTH, 14 JUNE 1977.

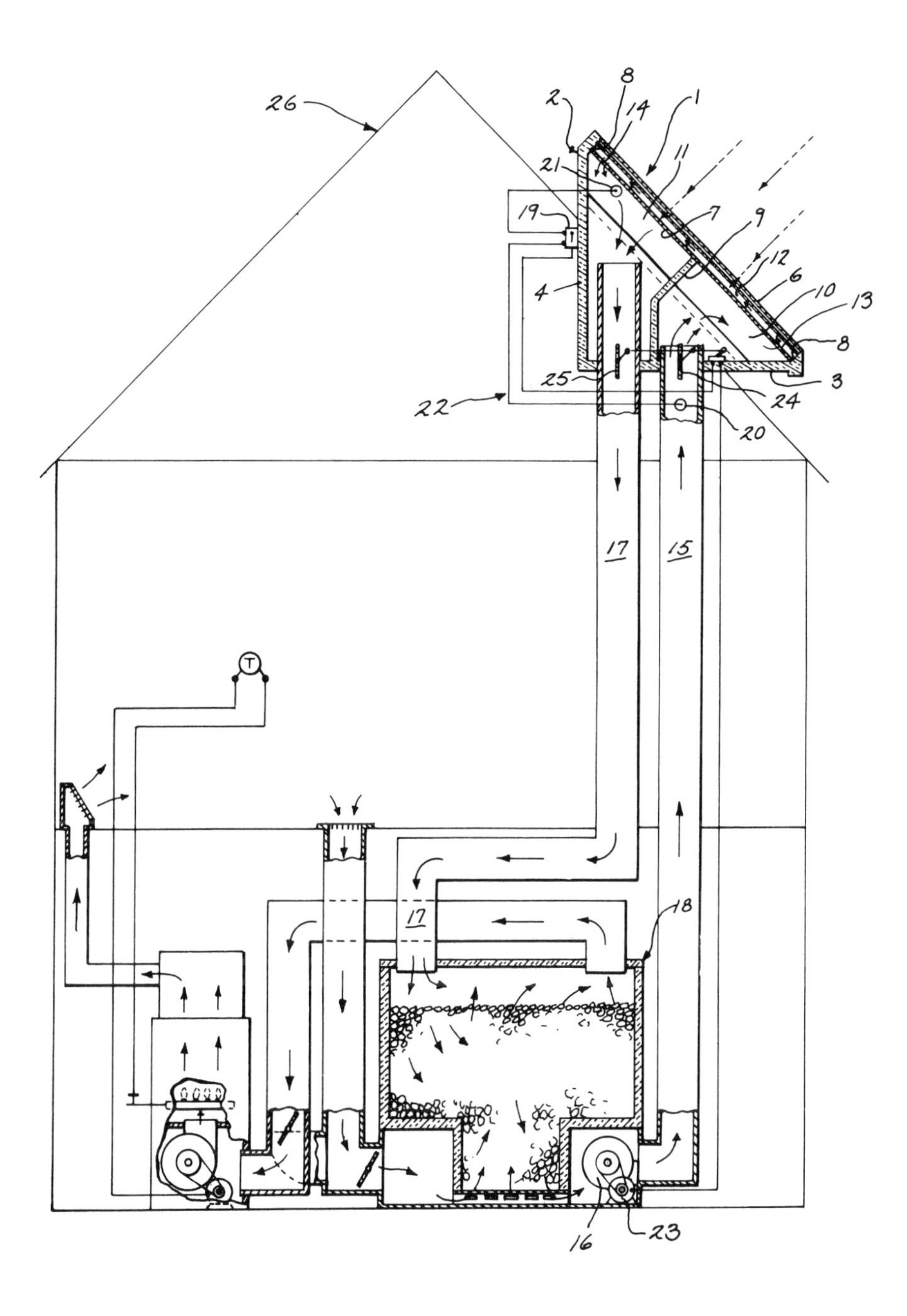

The Groth system illustrates what has come to be regarded as the "conventional" solar air heater for domestic use. Various modifications of this particular system have been patented by a number of inventors. Air is forced upward via duct (15) to collector (1) where it is heated and then returned to a stone-filled storage tank (18). When needed, the heated air in tank (18) may be pulled out by a furnace blower and distributed to a dwelling.

SEMICONDUCTOR ABSORBER FOR PHOTOTHERMAL CONVERTER

U.S. PATENT 4,037,014

ISSUED TO JONATHAN I. GITTLEMAN, 19 JULY 1977.

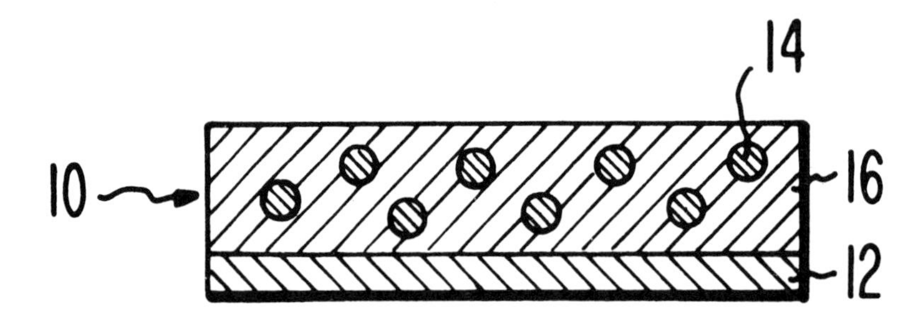

As any surface warms up, it begins to give off or "radiate" heat just as the sun radiates heat to us due to its extremely high temperature. Thus, one problem with solar collectors has been that, as they warm up, they tend to release heat back to the air before it can be captured and used. Gittleman uses semiconductor particles (14) in a solar absorber (16) to reduce this heat loss so that more of the sun's energy is contained.

SOLAR HEAT STORAGE SYSTEM

U.S. PATENT 4,037,652

ISSUED TO HANS BRUGGER, 26 JULY 1977.

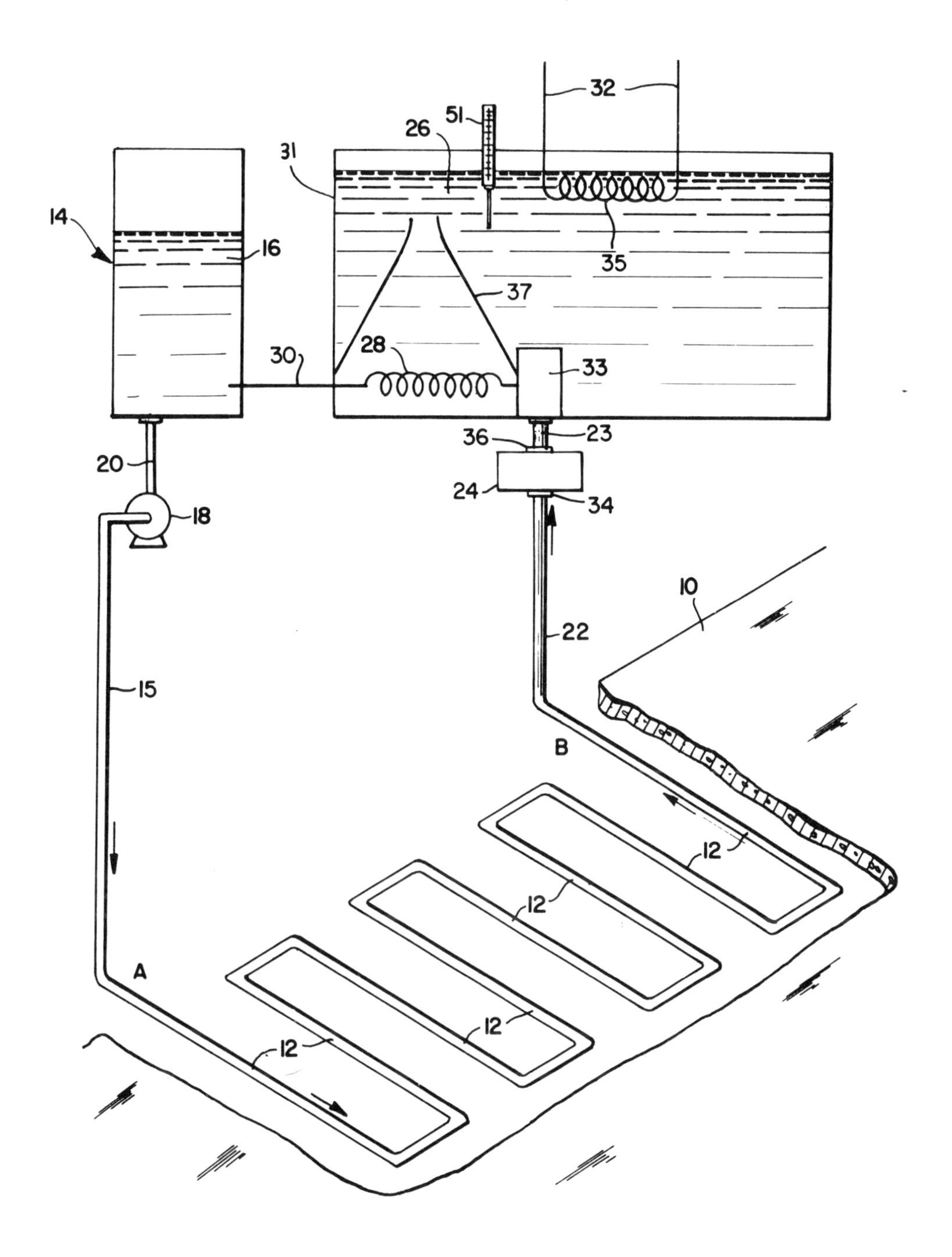

Our many blacktop roads and driveways are excellent heat absorbers and could serve as part of a solar-energy collector, as shown by this system. Brugger places collecting tubing (12) beneath blacktop pavement (10). The heated fluid in the tubes is pumped to tank (31) for immediate or later use. In winter, antifreeze (16) can be pumped through the tubes (12) to reduce icy road conditions.

POSITIONING A PLATFORM WITH RESPECT TO RAYS OF A LIGHT SOURCE

U.S. PATENT 4,041,307

ISSUED TO LOUIS S. NAPOLI ET AL., 9 AUGUST 1977.

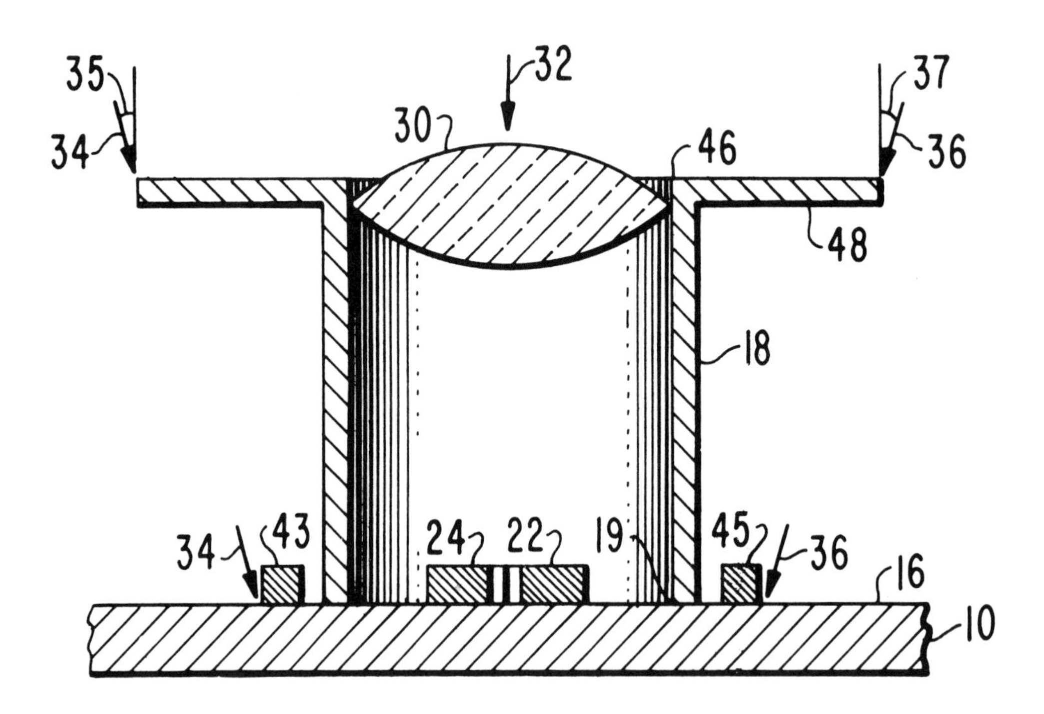

Many systems perform best when directly facing the sun, hence the need for sun "trackers." In Napoli's device, lens (30) concentrates sunlight on photocells (24) and (22) when it directly faces the sun. Cells (24) and (22) thus generate an electric current. Outer photocells (43) and (45) control movement of the device. When the sun is directly overhead, arms (48) shade cells (43) and (45) and they are not activated. When the sun is not directly overhead, one of cells (43) or (45) is struck by sunlight and sends an electric signal to move the collector to its optimal sun-facing position.

FIREPLACE/FORCED-AIR FURNACE HEAT GENERATION AND DISTRIBUTION SYSTEM

U.S. PATENT 4,049,194

ISSUED TO VERNON L. TICE ET AL., 20 SEPTEMBER 1977.

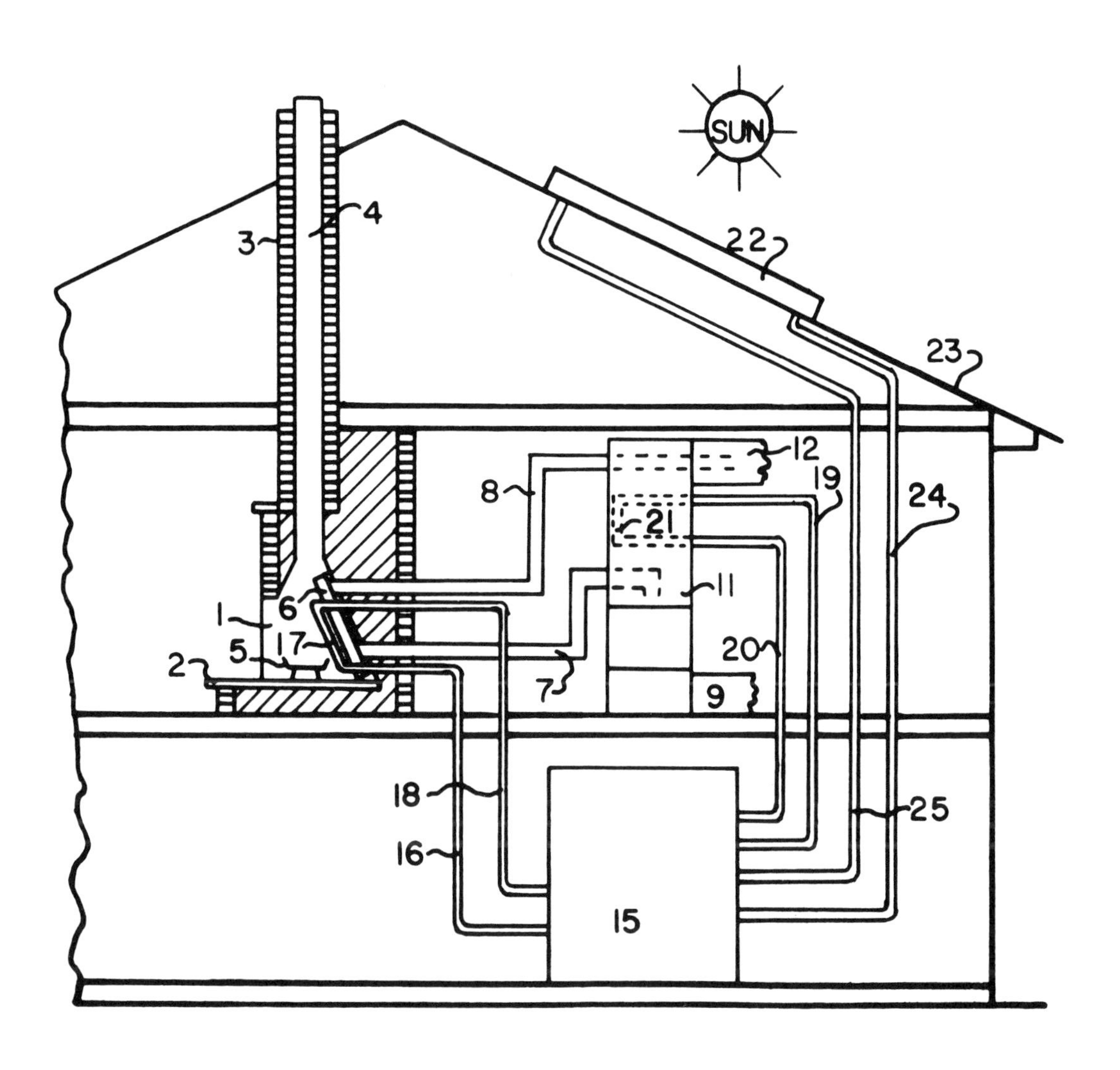

This invention combines a solar system and a fireplace heating system to produce a low-cost energy source. Fireplace heat from exchanger (17) and solar heat from collector (22) are stored in tank (15). When needed, this stored heat may be circulated through the home by means of the fan in central furnace (11).

SOLAR-ENERGY DEVICE AND SYSTEM

U.S. PATENT 4,050,443

ISSUED TO JOHN F. PECK ET AL., 27 SEPTEMBER 1977.

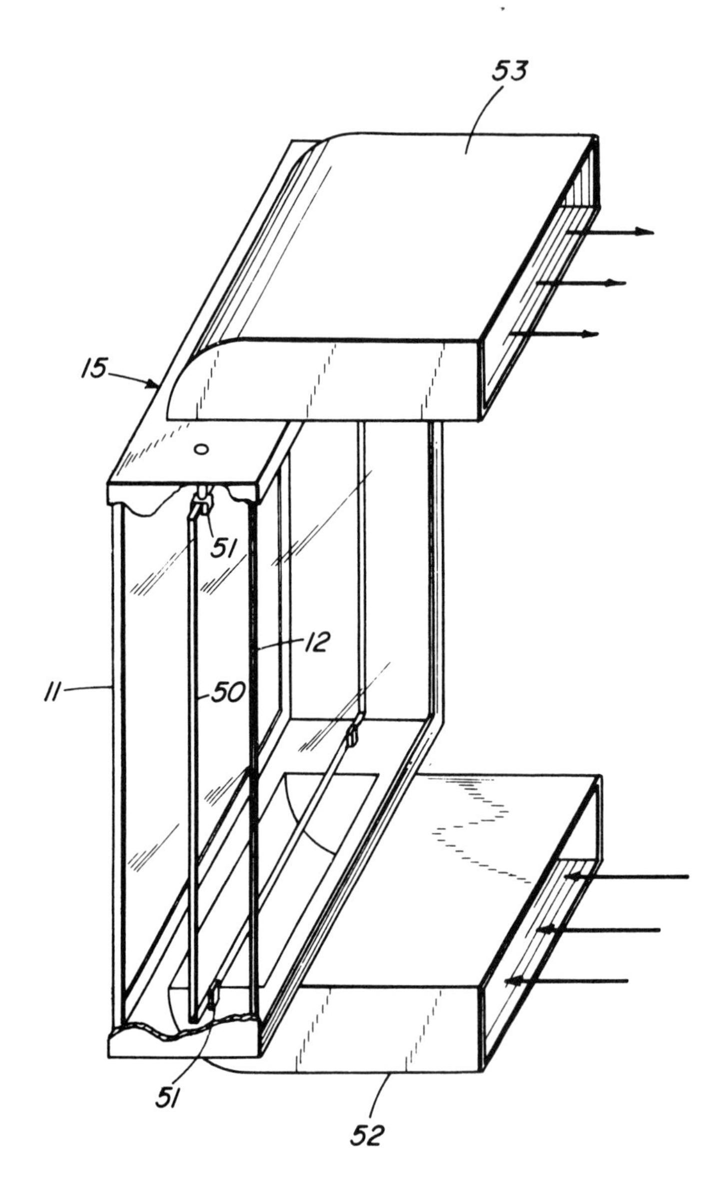

This patent illustrates the use of solar window units. Such units provide low-cost solar collectors that are easy to install and understand, thus helping the public to accept the use of solar energy in the home. The window unit consists of two clear panes (11) and (12) and pane (50), which is a darkened glass that partially transmits light and is heat absorbing. Circulating air is drawn into duct (52) and flows over both sides of pane (50) for optimal heat pick-up. The heated air rises to air distribution duct (53) and is transmitted into the room. An additional duct system can be hooked up to duct (53) to carry the heated air into areas that are remote from the sun side.

64

SOLAR-ENERGY REFLECTING APPARATUS WITH YIELDABLY BIASED REFLECTORS

U.S. PATENT 4,050,777

ISSUED TO MELVILLE F. PETERS, 27 SEPTEMBER 1977.

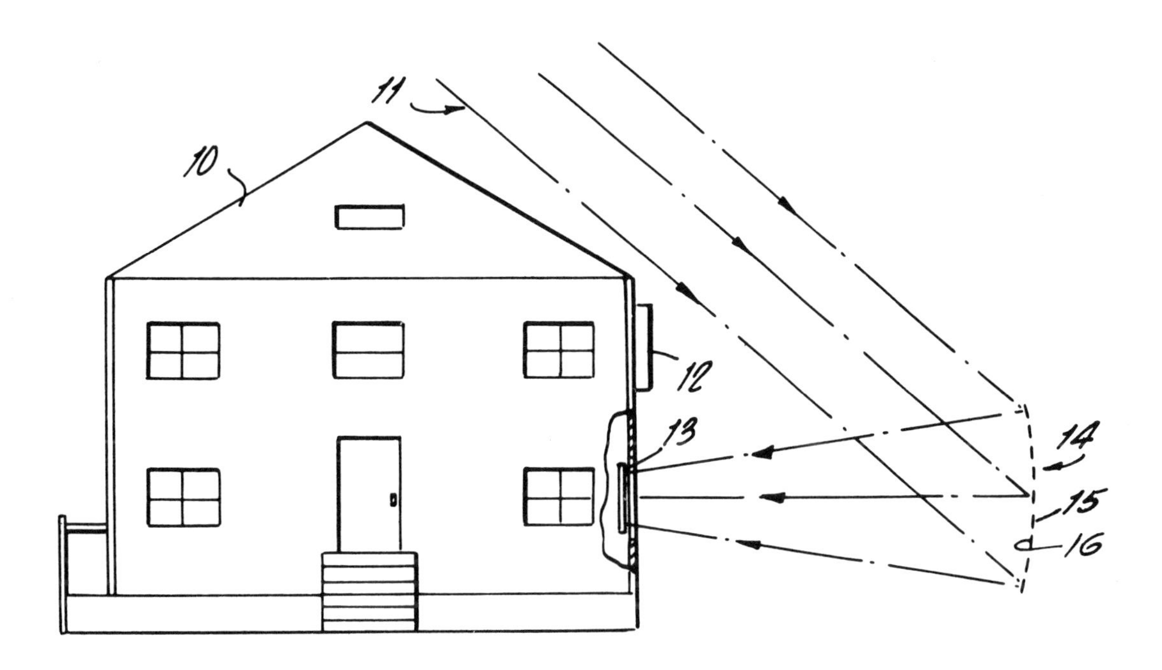

The north side of a home is always cooler, since it receives no direct sunlight. Peters uses mirrors (14) outside the house to reflect winter sun into the north rooms. Since they are exposed to the elements, the mirrors (14) are designed to bend under high wind conditions to avoid breakage. One advantage of this idea is that no modification need be made to the structure itself.

SOLAR HEATING PIPE

U.S. PATENT 4,051,835

ISSUED TO GEORGE HINSON-RIDER, 4 OCTOBER 1977.

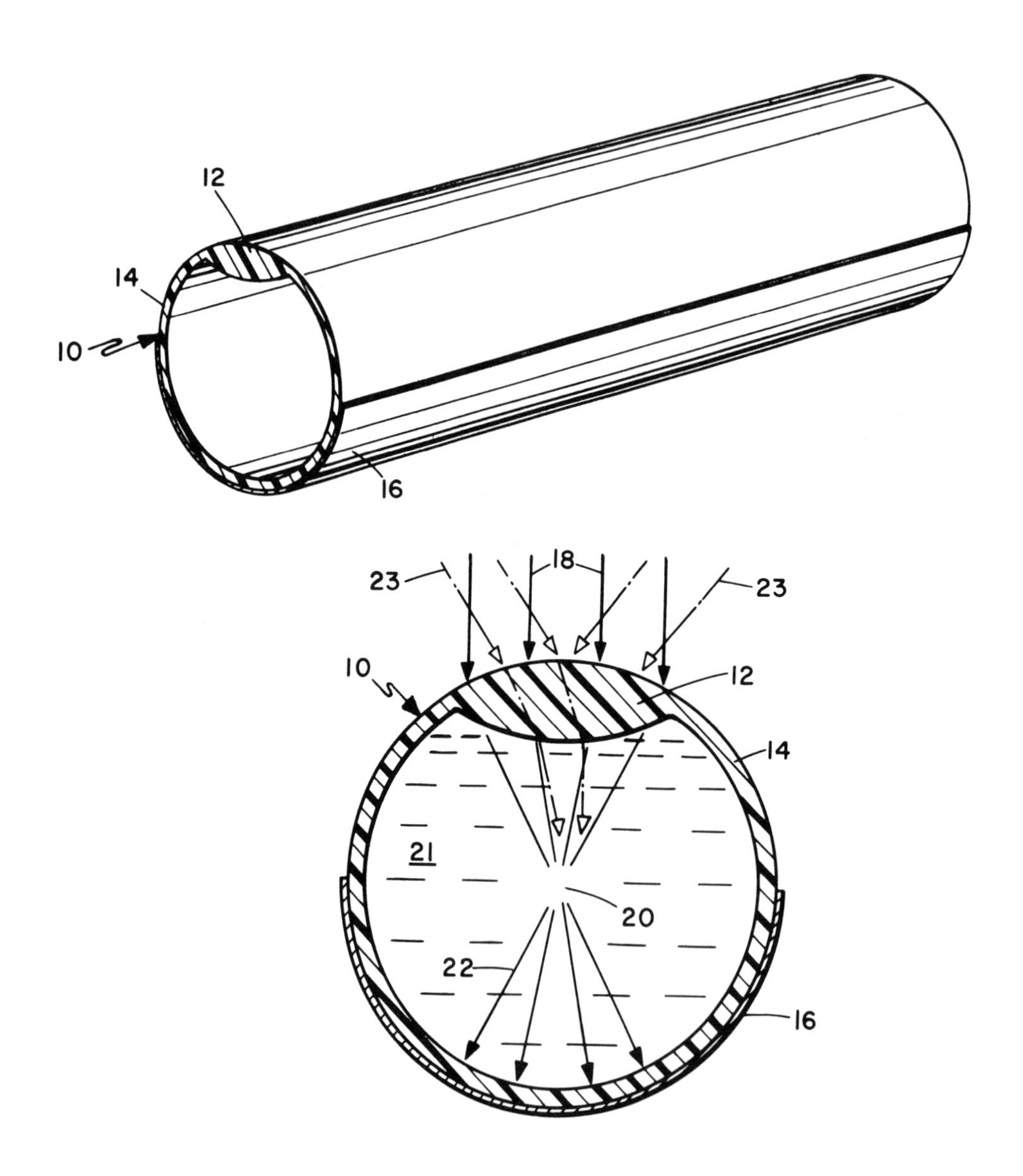

By combining the parts of a solar collector into a single, pipelike piece, this device could reduce installation and maintenance costs. Sun-concentrating lens (12) is formed as a part of plastic or glass tube (10). Thus, liquid flowing through the tube is more efficiently heated. An absorptive coating (16) may be applied to the lower portion of the tube to reduce escape of concentrated solar rays from it.

METHANE-GAS PROCESS AND APPARATUS

U.S. PATENT 4,057,401

ISSUED TO OLIVER W. BOBLITZ, 8 NOVEMBER 1977.

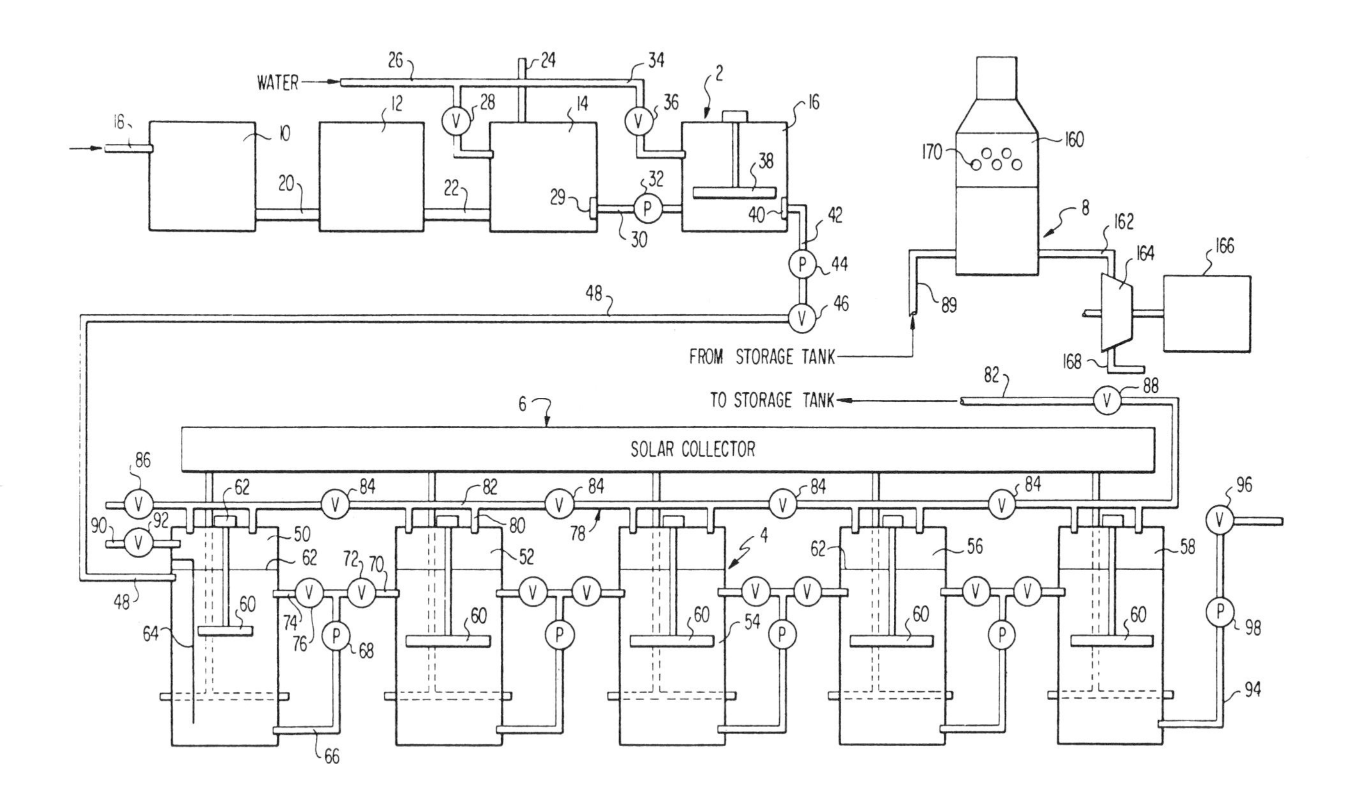

Methane gas, a very clean-burning fuel, is produced by allowing organic wastes to decay in a tank. This device uses solar energy to heat the tank and thus speed up the gas-producing process. Organic wastes are fed into pipe (18) and prepared and separated in chambers (10) to (16), then fed into digester tanks (50) to (58) via pipe (48). Instead of other costly fuels, solar collector (6) is used to heat the tanks. The methane gas produced flows in pipe (82) to a storage tank and, when needed, is fed in pipe (89) to burner (8). The invention suggests that methane-gas generating plants be located in areas of high sun concentration to reduce digester tank heat-energy costs.

SOLAR HEATING AND INSULATING APPARATUS

U.S. PATENT 4,058,109

ISSUED TO RONALD J. GRAMM, 15 NOVEMBER 1977.

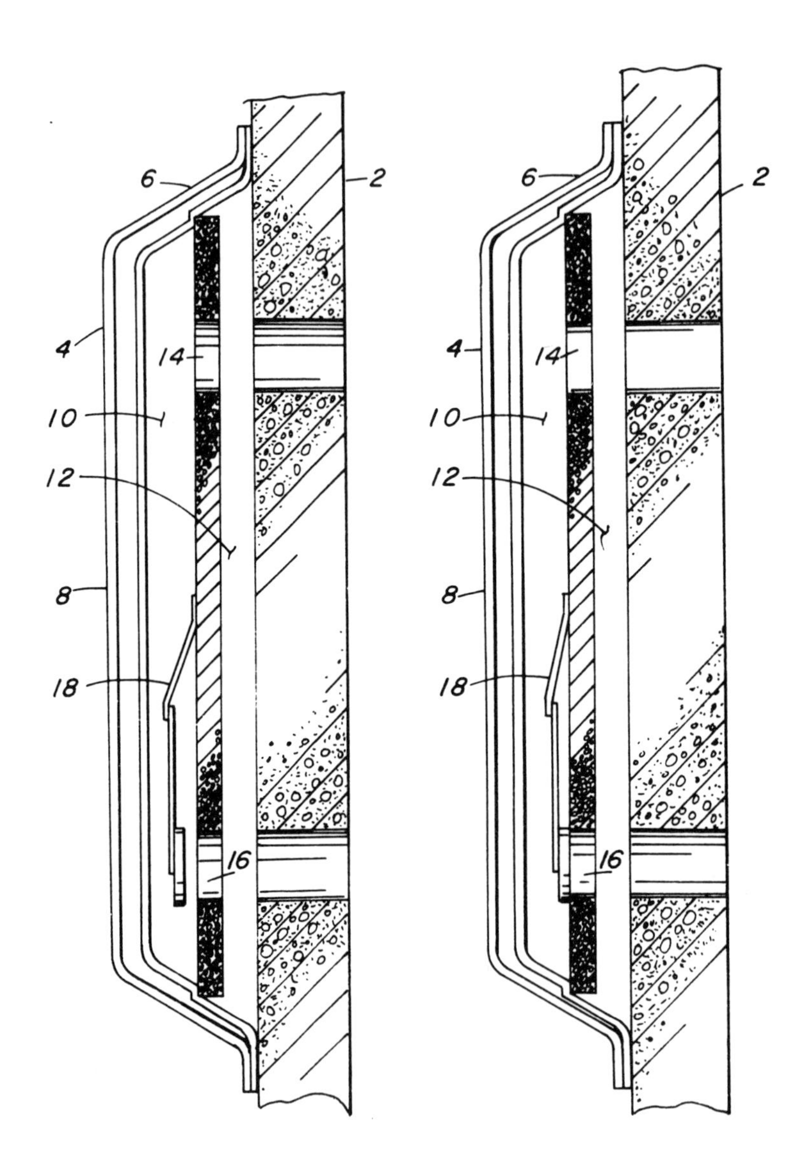

Many older buildings were not designed with energy-saving in mind. This device is used to convert windowless but sun-facing building walls into solar collectors. Transparent piece (4) is attached to windowless wall (2). A heat absorbing plate has ducts (14) and (16), and two small ducts are formed in the windowless wall corresponding to them. When the sun heats air in chamber (10), a bimetallic arm (18) opens the valve covering duct (16) to allow circulation of inside building air through duct (16) upward to duct (14) and back into the building.

68

DEVICE FOR UTILIZING THE HEAT ENERGY OF SOLAR RADIATION

U.S. PATENT 4,059,095

ISSUED TO EDGARD GRUNDMANN ET AL., 22 NOVEMBER 1977.

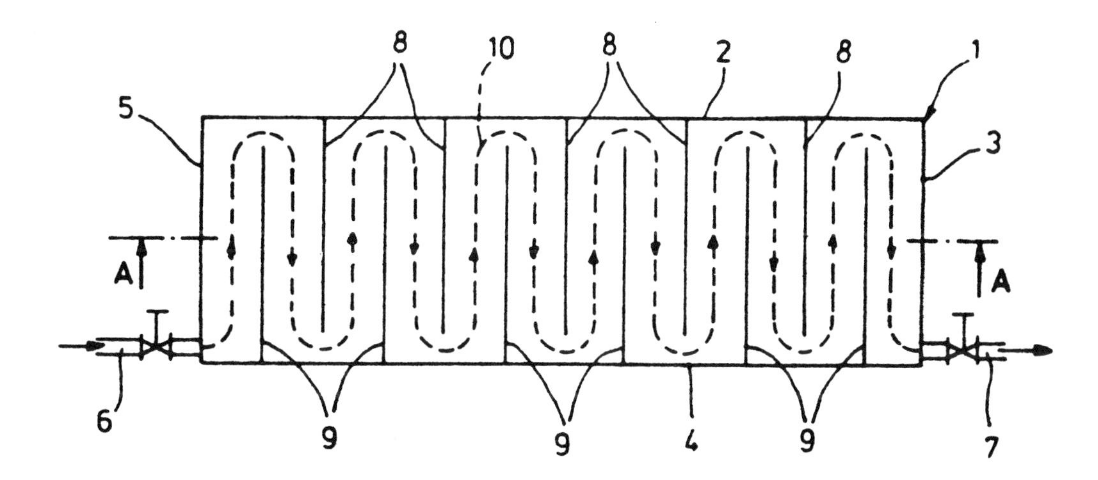

Because of its portable features, this device suggests the use of solar collectors in temporary locations such as camping or construction sites. In use, air flows through the clear plastic conduits (8) and is heated by the sun. When no longer required, the lightweight collector can be easily rolled up and moved.

HEAT COLLECTOR AND STORAGE CHAMBER

U.S. PATENT 4,059,226

ISSUED TO DAVID L. ATKINSON, 22 NOVEMBER 1977.

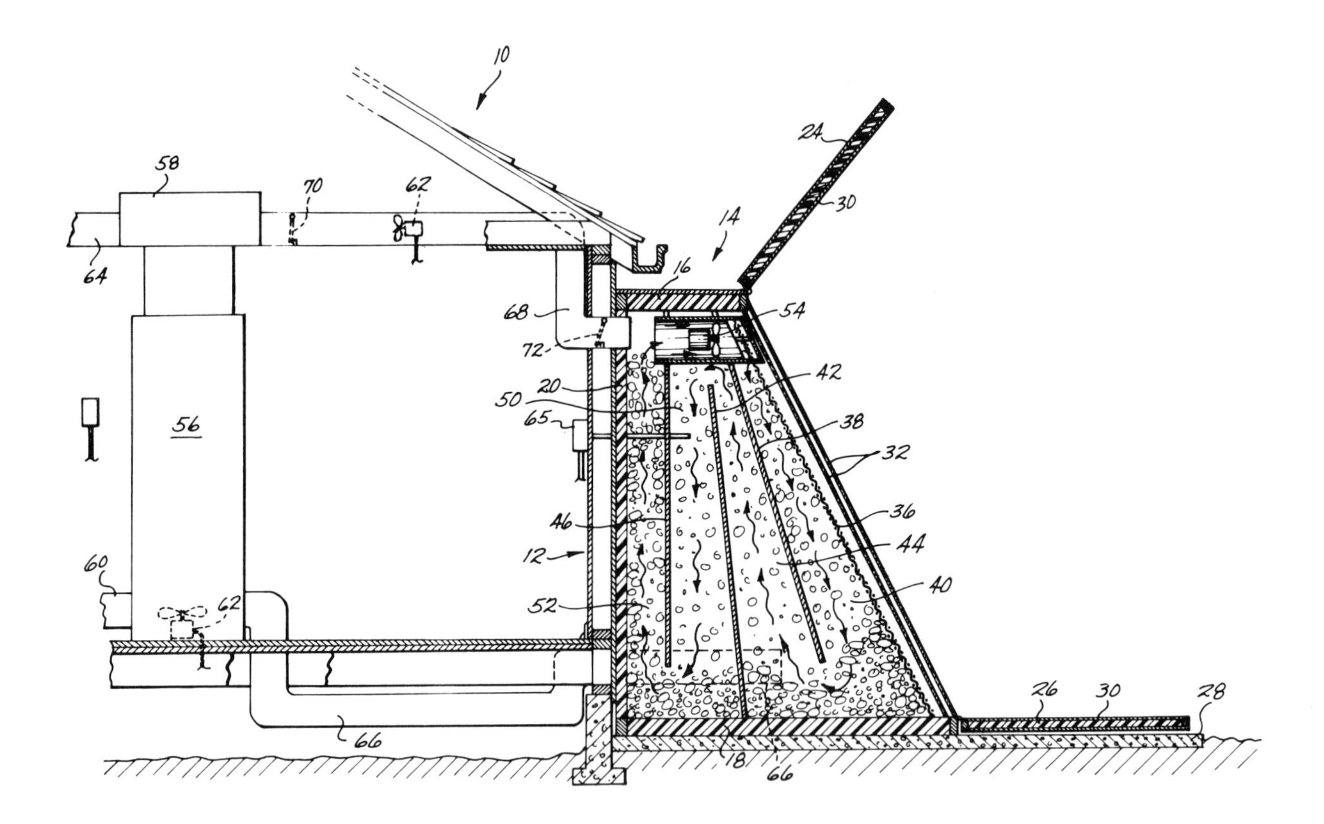

Atkinson's design shows an inexpensive heating system that can be added to the side of a house without making major changes to the house structure. Glass (32) allows solar rays to pass into loosely packed storage rocks (40), which are held in place by screens (36). Baffles (38), (42), and (46) direct air flow through the rocks so that heat stored in them can be moved into the home. Doors (24) and (26) are closed at night to prevent heat loss.

SOLAR-HEAT EDUCATIONAL DEVICE

U.S. PATENT 4,060,916

ISSUED TO GEORGE D. FINIGAN, 6 DECEMBER 1977.

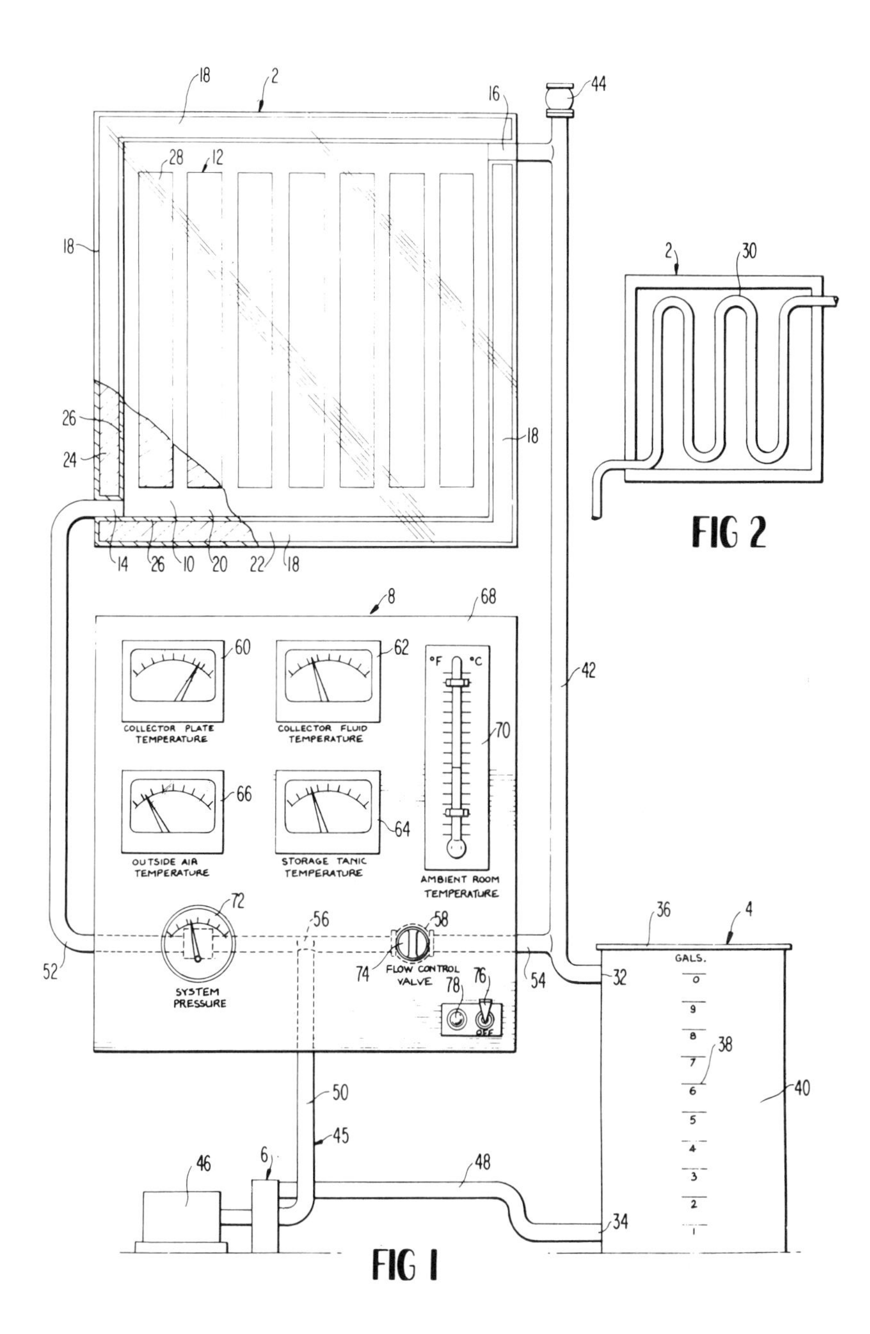

Teaching people how well solar heaters work is an important first step in having them accepted as a primary energy source. In this educational system, solar collector (2), storage tank (4), and the piping between them are made of transparent materials wherever possible. This allows maximum viewing of sun-heated liquid so that flow and boiling patterns can be easily understood. Indicator panel (8) shows system temperature and pressure readings at critical locations.

SOLAR REACTOR COMBUSTION CHAMBER

U.S. PATENT 4,070,861

ISSUED TO ROBERT L. SCRAGG ET AL., 31 JANUARY 1978.

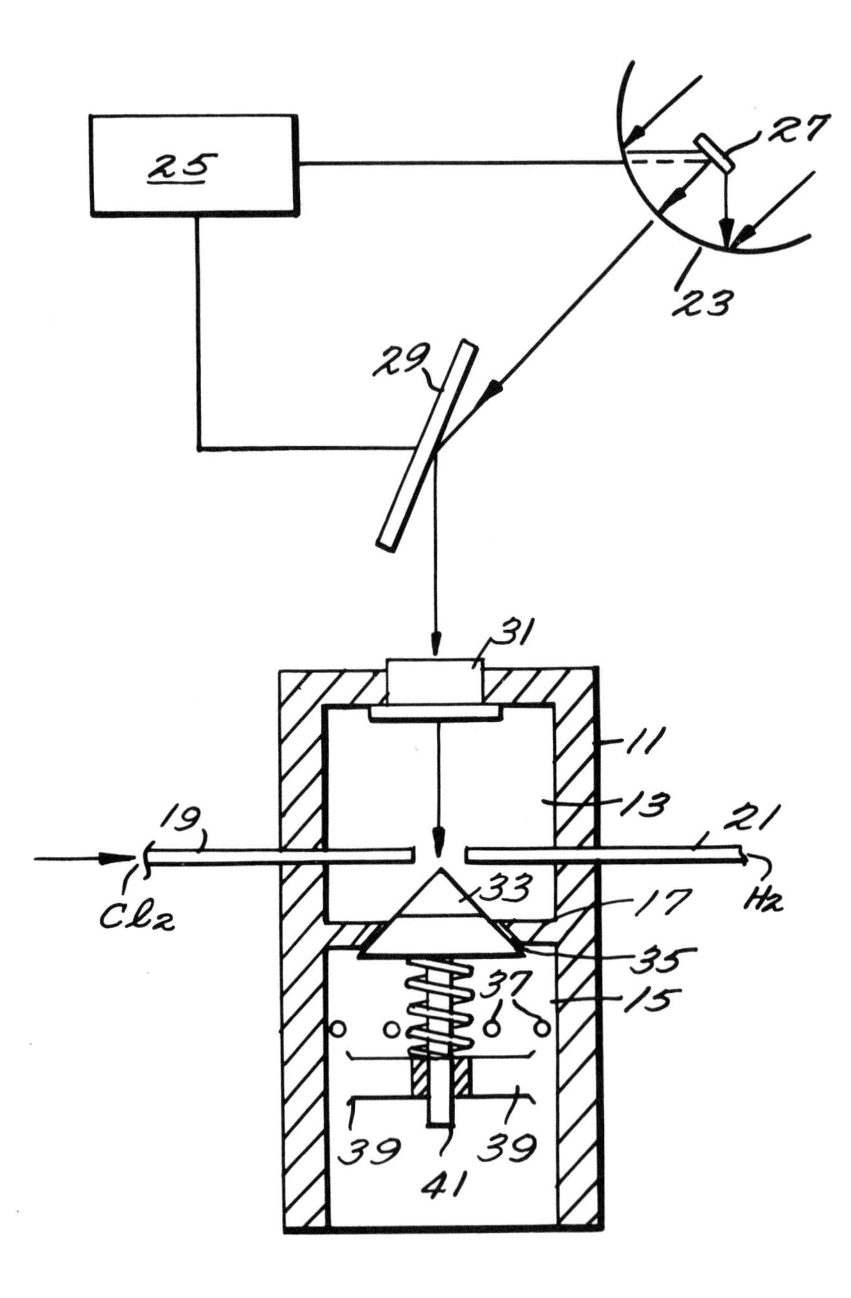

Solar energy can supply heat to cause two or more chemicals to react with each other. In this device, concentrated sunlight is reflected at (29) into chamber (13). Hydrogen and chlorine are fed into the chamber and, in the presence of the sun's heat energy, form hydrogen chloride gas, which expands past valve (33) into the lower combustion chamber (15). There oxygen is introduced at (37). The hydrogen chloride and oxygen react with an explosive force, which may be used to drive an engine or turbine.

SOLAR CIGARETTE LIGHTER

U.S. PATENT 4,076,014

ISSUED TO A. W. WIQUEL, 28 FEBRUARY 1978.

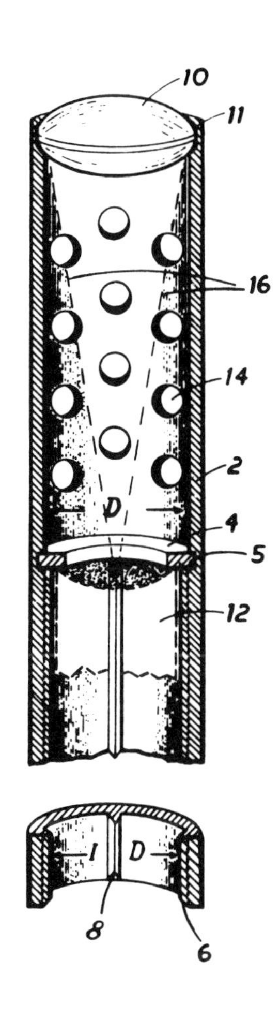

Although the surgeon general may not be overjoyed with this particular invention, the solar cigarette lighter shows one of the many ways in which concentrated sunlight can be used to ignite something. In operation, a cigarette (12) is inserted into tube (6) as far as ring (5). Solar rays are concentrated by lens (10) to ignite the end of the cigarette. When lit, smoke comes out at holes (14) and the cigarette is withdrawn. Since the solar cigarette lighter is not affected by wind, it is very useful at beach, boating, and skiing areas.

APPARATUS FOR THE UTILIZATION OF SOLAR HEAT

U.S. PATENT 4,076,016

ISSUED TO MAURICE E. PHILLIPS, 28 FEBRUARY 1978.

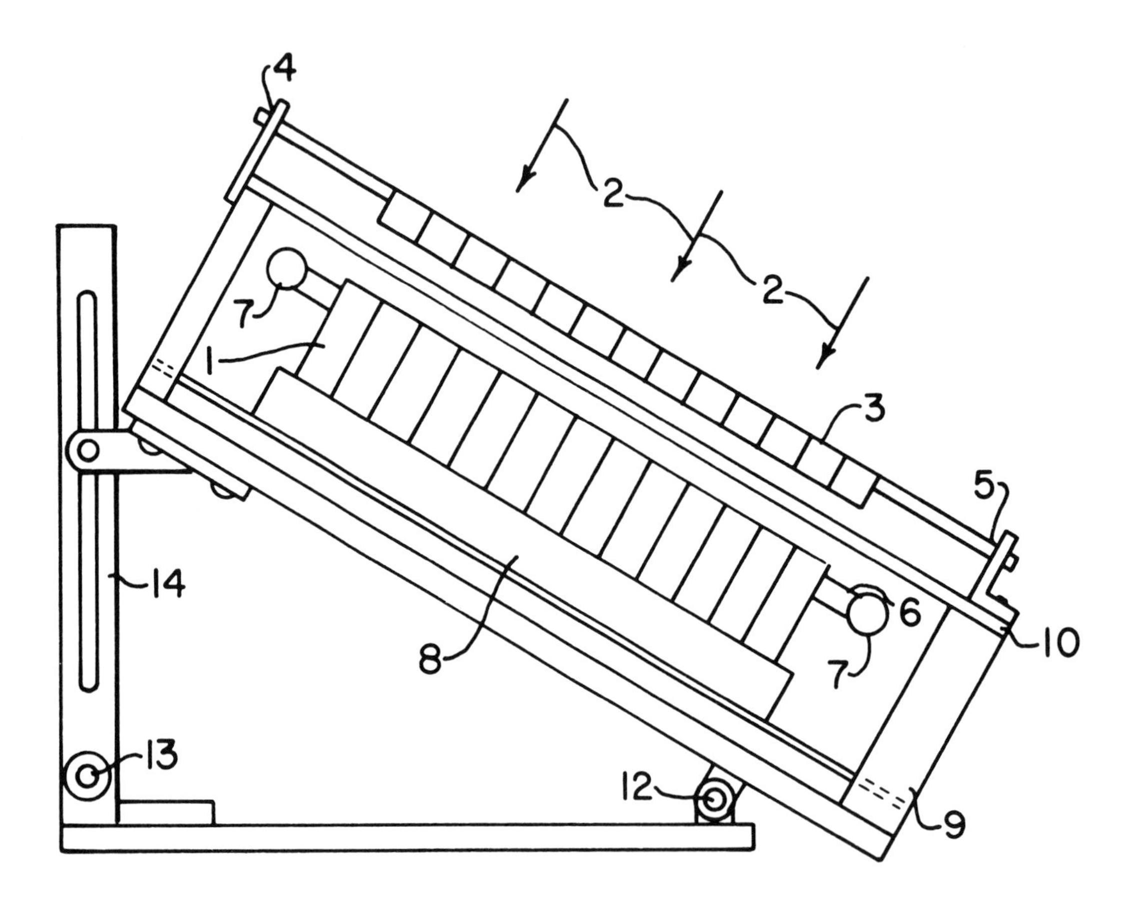

The Phillips patent shows a simple way to rotate a solar collector so that it directly faces the sun. In use, collector (8) is pivoted about point (12) and held at the required angle by a pin in slot (14). Thus, a single mass-produced collector (8) could be shipped to any geographic area and be easily adjusted to the sun angle present there. Also, as the sun changes with the seasons in a particular location, a simple manual adjustment could be made.

SOLAR-ENERGY-OPERATED MOTOR APPARATUS

U.S. PATENT 4,079,249

ISSUED TO KENNETH P. GLYNN, 14 MARCH 1978.

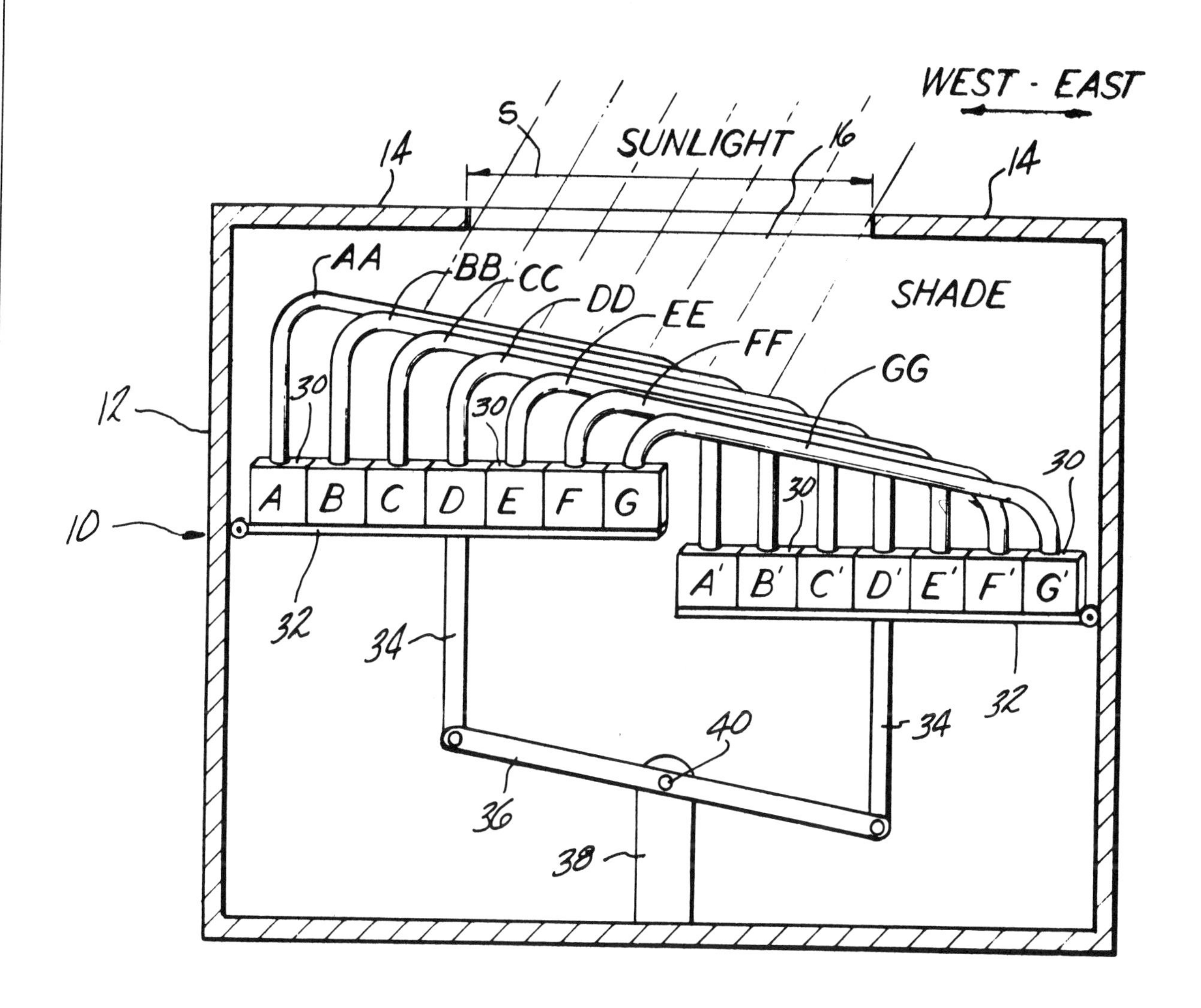

Many solar collectors require a gas or electric motor to turn them toward the sun, i.e., to "track" the sun. In this device, solar energy drives the tracking motor, thus saving money on gas or electricity. In operation, a fluid in chambers (A), (B), (C), etc., is partially boiled and flows through lines (AA), (BB) to right chambers (A′), (B′), etc. This rotates arm (36), which turns a solar collector to face the sun. In the morning, containers (A′), (B′), etc., are shaded so no boiling occurs. As the sun angle progresses to the west, more fluid in the right chambers is boiled and transferred to the left chambers, thus rotating arm (36).

PORTABLE SOLAR-ENERGY COLLECTOR

U.S. PATENT 4,080,955

ISSUED TO WAYNE R. SANDSTROM, 28 MARCH 1978.

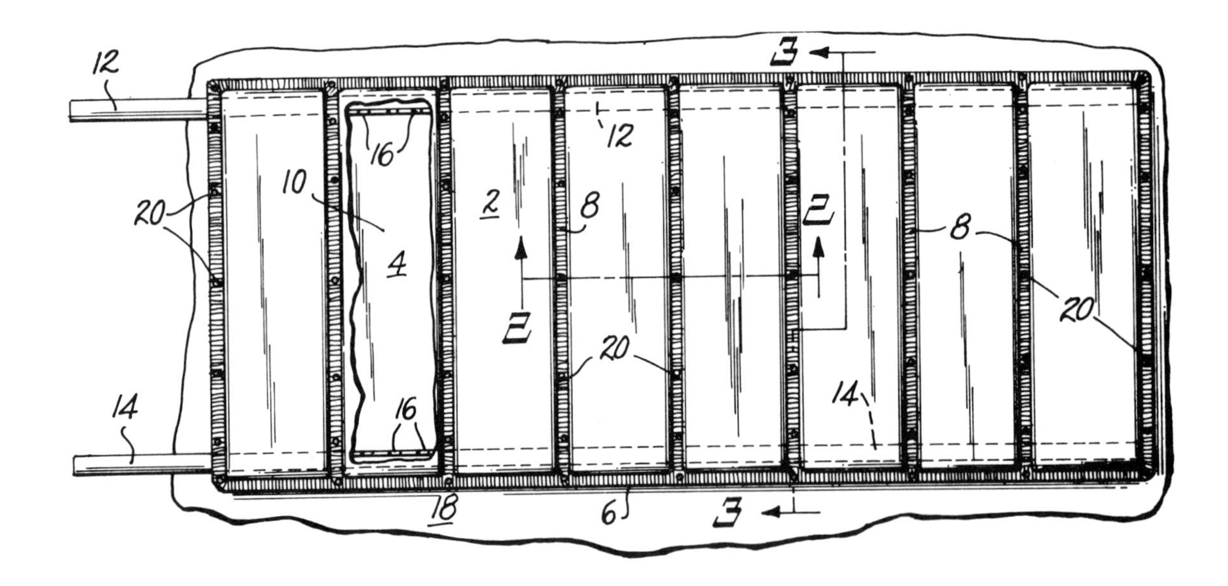

A fluid can be heated more quickly by the sun if it is divided into smaller amounts. Sandstrom's device comprises a flexible bag with multiple compartments separated by dividers (8). Fluid enters and exits the collector via tubes (12) and (14). Each compartment has its own water distribution holes (16) so that there is a uniform flow of a small amount of liquid to each section.

PORTABLE SOLAR OVEN AND GRILL

U.S. PATENT 4,082,079

ISSUED TO JOHN RODGERS, 4 APRIL 1978.

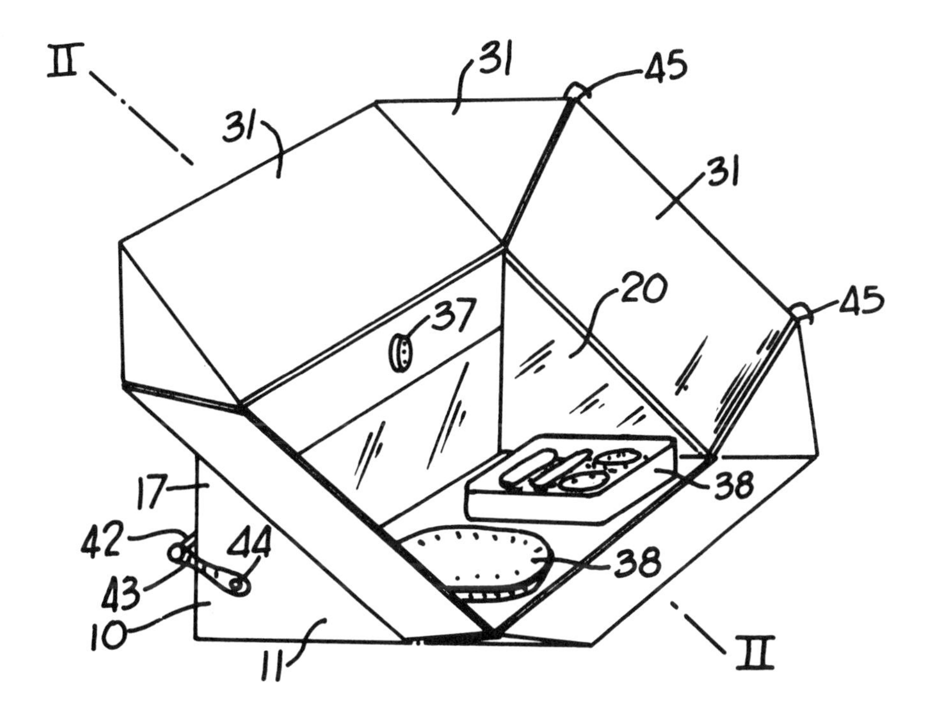

The use of solar ovens could eliminate the pollution and danger of the traditional summer barbecue. As shown by Rodgers, food items (38) are placed in the oven and are covered by transparent layer (20), which admits solar rays to heat the food and also seals the interior from atmospheric influences. The interior walls of the oven are coated black to absorb solar energy, while outer panels (31) are reflective and angled to direct sunlight toward the interior. This particular invention is foldable and portable via handle (42). The use of solar cooking ovens illustrates a relatively high-temperature application of solar energy on a small scale.

SOLAR ENERGY

U.S. PATENT 4,082,143

ISSUED TO HARRY E. THOMASON, 4 APRIL 1978.

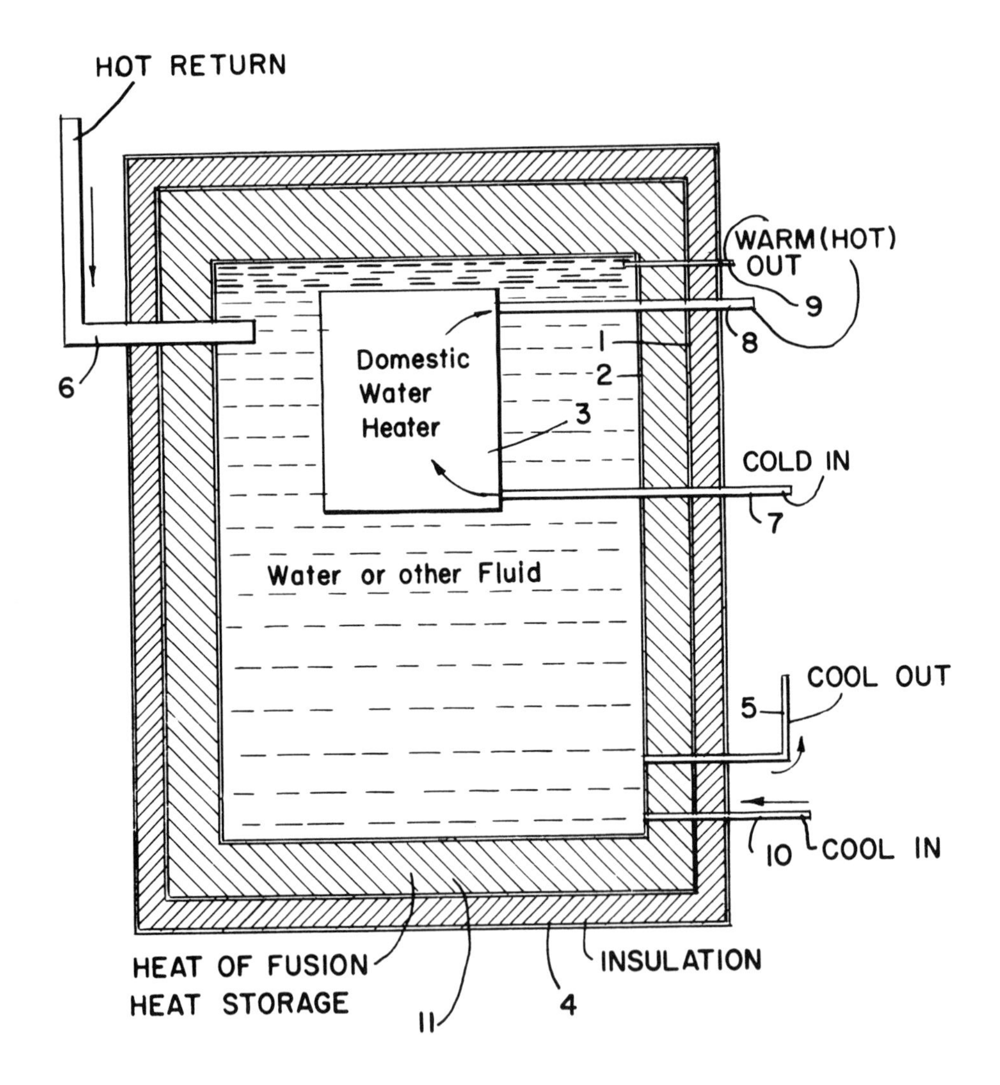

Water for showers or other household uses may be heated by solar energy instead of by costly gas or electric water heaters. In the Thomason system, solar-heated water in pipe (6) enters an insulated storage tank (4). A container for household water (3) is positioned in the warmer upper part of the storage tank to be heated and water is pumped out at pipe (8) for domestic use. Since the domestic water is not itself pumped through a solar collector, it remains clean and uncontaminated.

DEVICE FOR COLLECTING SOLAR ENERGY

U.S. PATENT 4,083,360

ISSUED TO J. F. COURVOISIER ET AL., 11 APRIL 1978.

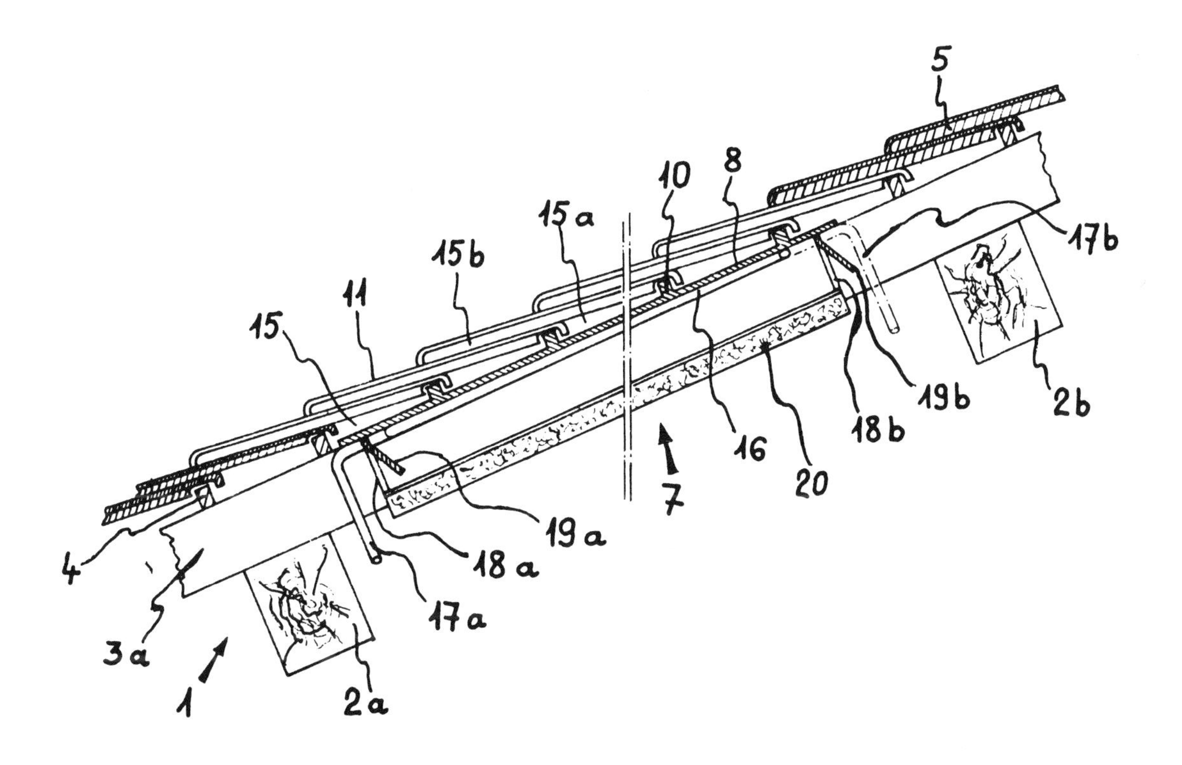

By having a solar collector built into the roof, this system does not detract from a home's appearance. In use, transparent roofing tiles (11) allow rays of sun to pass through them, strike an absorber plate (10), and heat air flowing through ducts (18a) and (18b). Since the tiles overlap, insulating air spaces are formed at (15b) to help retain the sun's heat. Water pipes (16) may also be added to capture more sun energy.

ATTIC EXHAUST SYSTEM

U.S. PATENT 4,085,667

ISSUED TO NEIL B. CHRISTIANSON, 25 APRIL 1978.

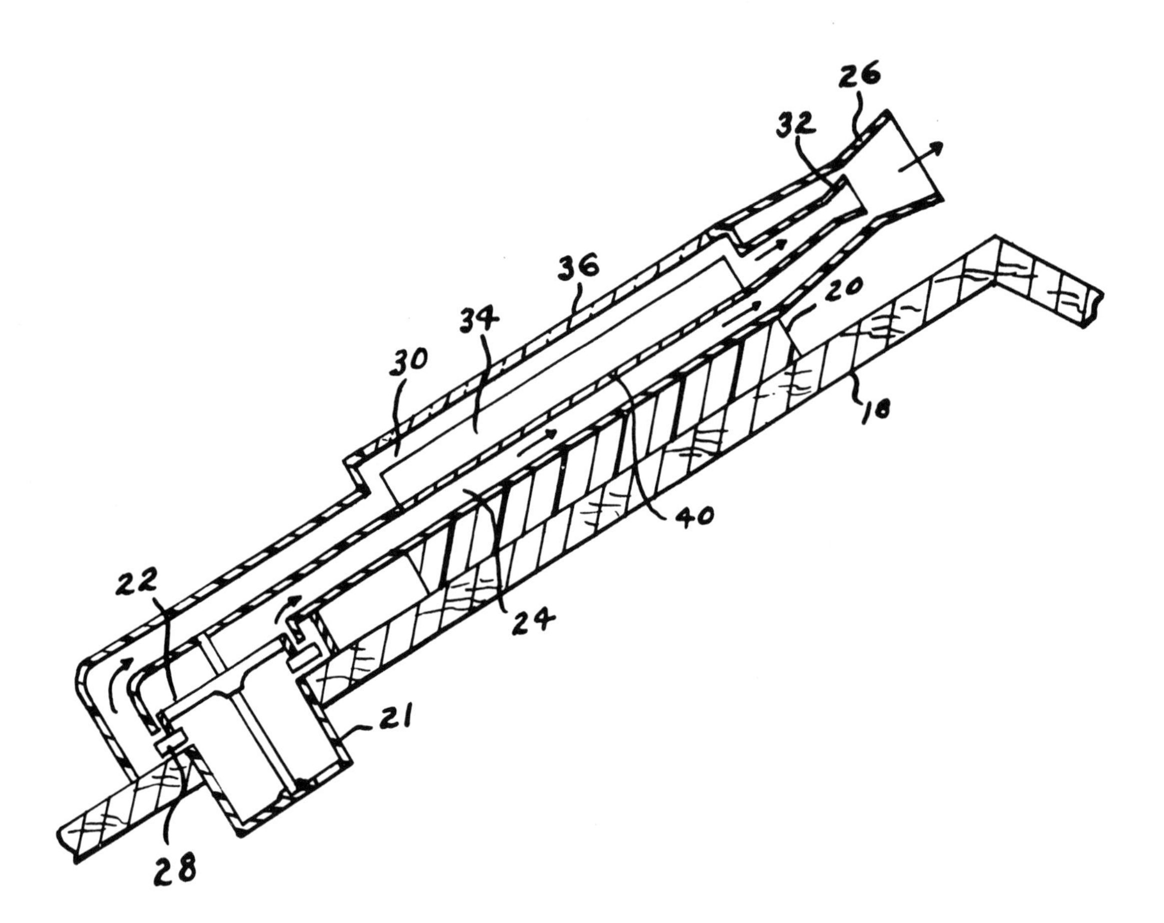

A home attic can be ventilated by using solar power instead of a conventional electric fan. In operation, a clear plate (36) allows the solar heating of passage (30), thus creating an upward flow of air, which passes through nozzle (32). The flow out of nozzle (32) has the effect of drawing air through duct (24) to vent the attic via opening (21). Fortunately, the device functions best during the hot summer months, when it is most needed.

SOLAR-HEAT COLLECTOR CONSTRUCTION

U.S. PATENT 4,086,913

ISSUED TO JOSEPH G. GAVIN, 2 MAY 1978.

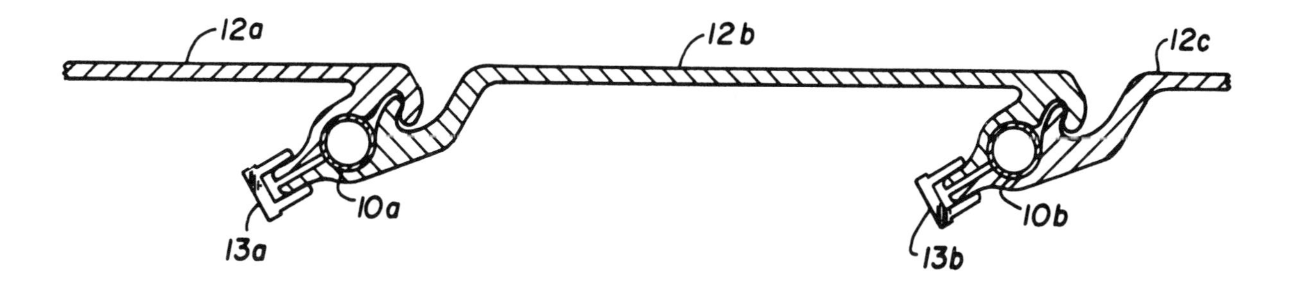

Plate-and-tube collectors, those with a sun-absorbing plate and fluid-carrying tubing, have found widespread use in the solar field. This patent shows such a collector, which can be quickly constructed or disassembled for repair, thus reducing labor costs. In use, sun-absorbing plates (12a, 12b, and 12c) have ends shaped so they can be fitted to surround tubes (10) for optimal heat transfer to the fluid in them. The plates are then easily fastened together by clips (13).

RADIANT ENERGY COLLECTOR

U.S. PATENT 4,088,116

ISSUED TO JOSE PASTOR, 9 MAY 1978.

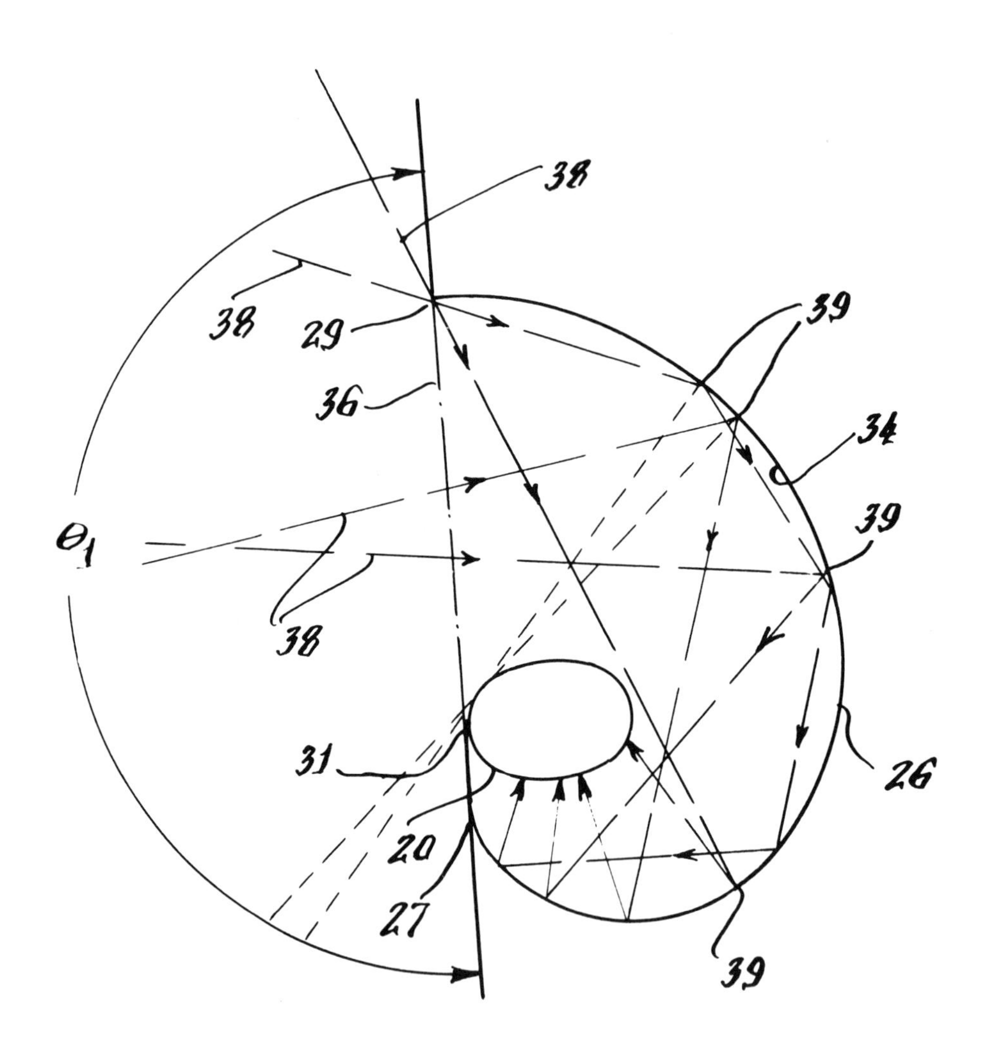

This invention shows the use of a stationary reflector shaped like a scroll. The scroll design enables the reflector to capture sunlight from various angles (38) and direct it onto a sun collector. This eliminates the need for a moving sun-tracking system. In practice, solar rays hit the scroll-shaped reflector (34) and are directed onto a fluid-carrying tube (20).

COMPOUNDS FOR FORMING BODIES FOR GENERATING HEAT FROM RADIANT LUMINOUS ENERGY

U.S. PATENT 4,090,496

ISSUED TO GILBERT MALLET, 23 MAY 1978.

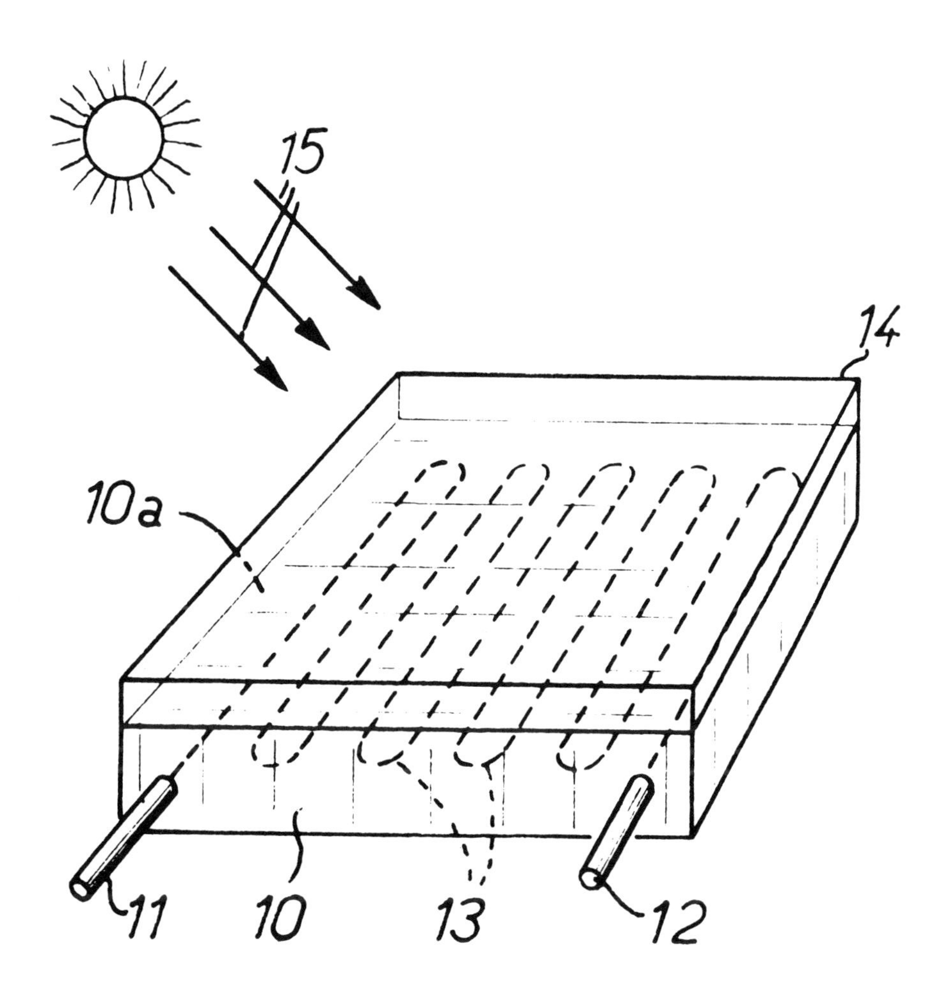

Solid plastic or resin solar collectors have the advantages of being long lasting and less subject to breakage during installation and use. As in the mold casting of plastics, etc., liquid-carrying tubes (13) are placed in a mold and a clear resin mixture (10) is poured around them to form a rectangular collector. To make the collector more light-absorbent, black particles may be dispersed throughout the resin mix. After the resin mixture has solidified, it is removed from the mold and a transparent cover (14) is added to provide an insulating effect.

SOLAR POND

U.S. PATENT 4,091,800

ISSUED TO JAMES C. FLETCHER, 20 MAY 1978.

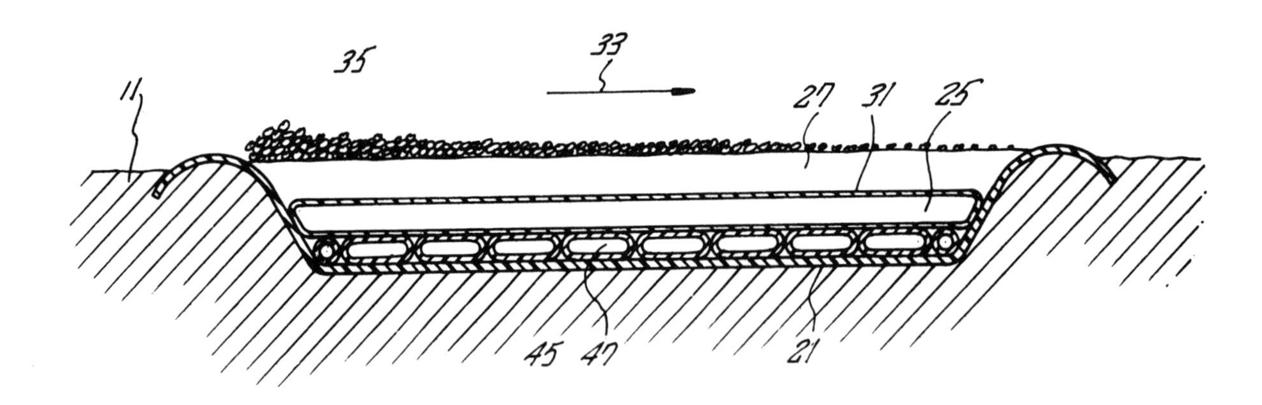

This system shows the use of long bulldozed trenches that may be dug to form water ponds for absorbing solar energy. Heat-absorbing liquid layers (25) and (47) are sealed in plastic bags to keep them clean. Lower plastic liner (21) may be painted black to increase absorption of sunlight. An upper liquid cover layer (27) insulates the lower plastic bags to reduce heat losses into the atmosphere.

SOLAR ENGINE

U.S. PATENT 4,094,146

ISSUED TO EARL O. SCHWEITZER, 13 JUNE 1978.

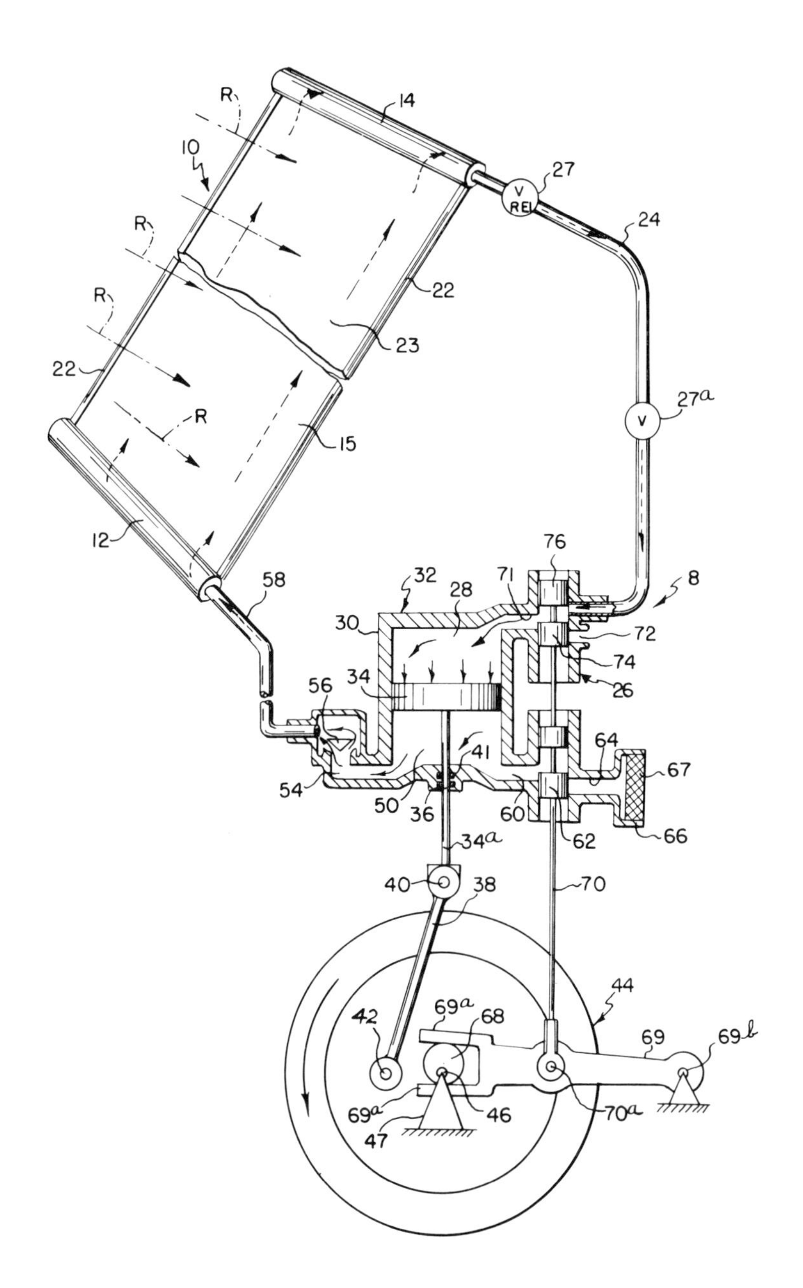

This device shows another type of engine driven by solar energy. Solar-heated air from collector (23) flows through piping (24) to piston chamber (28). Piston (34) is thus forced downward to drive crank (42) and wheel (44). Rod (70), which is mounted on driven wheel (44), controls air inlet valves (74) and (62). The upper valve (74) allows entry of solar-heated air while lower valve (62) allows entry of cooler atmospheric air. As piston (34) descends, the cooler air is driven through tube (58) to be heated in collector (23) and the process is repeated.

SOLAR HEAT COLLECTOR

U.S. PATENT 4,094,300

ISSUED TO SAM W. YOUNG, 13 JUNE 1978.

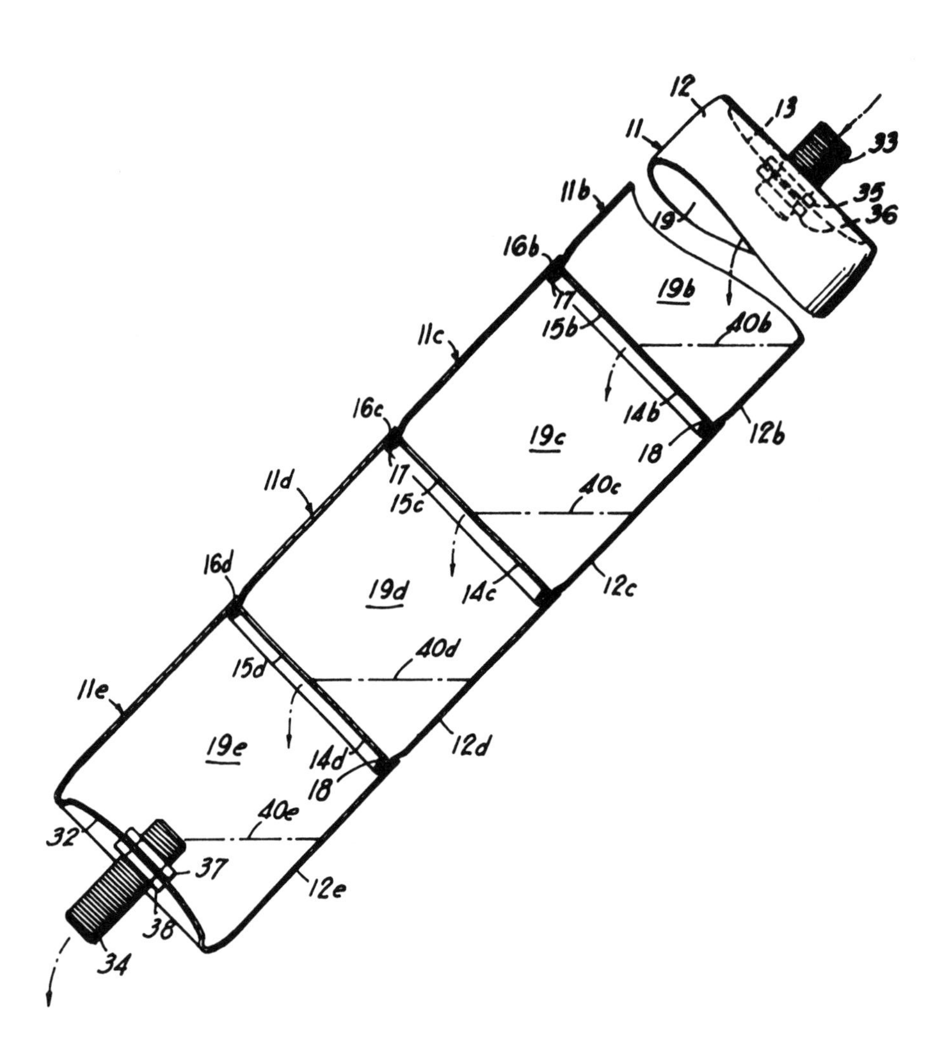

Used aluminum beverage cans, which would normally be thrown away or recycled, can be made into an efficient solar collector. After the bottom ends have been removed from the beverage cans (11), the cans are welded together, so that their openings are at upper locations (15). As liquid entering at (33) flows down through the slanted collector, it is delayed at points (40b, 40c, 40d, and 40e). This lengthens the time the fluid is subject to the heated air within the chambers, increasing the heating effect. Heated fluid then overflows to the next can and finally out the exit tube (34) to be used. Aluminum cans are especially well suited for this system because of their good heat-transfer properties.

SOLAR HEATING ASSEMBLY

U.S. PATENT 4,102,324

ISSUED TO OLIVER E. NETHERTON, 25 JULY 1978.

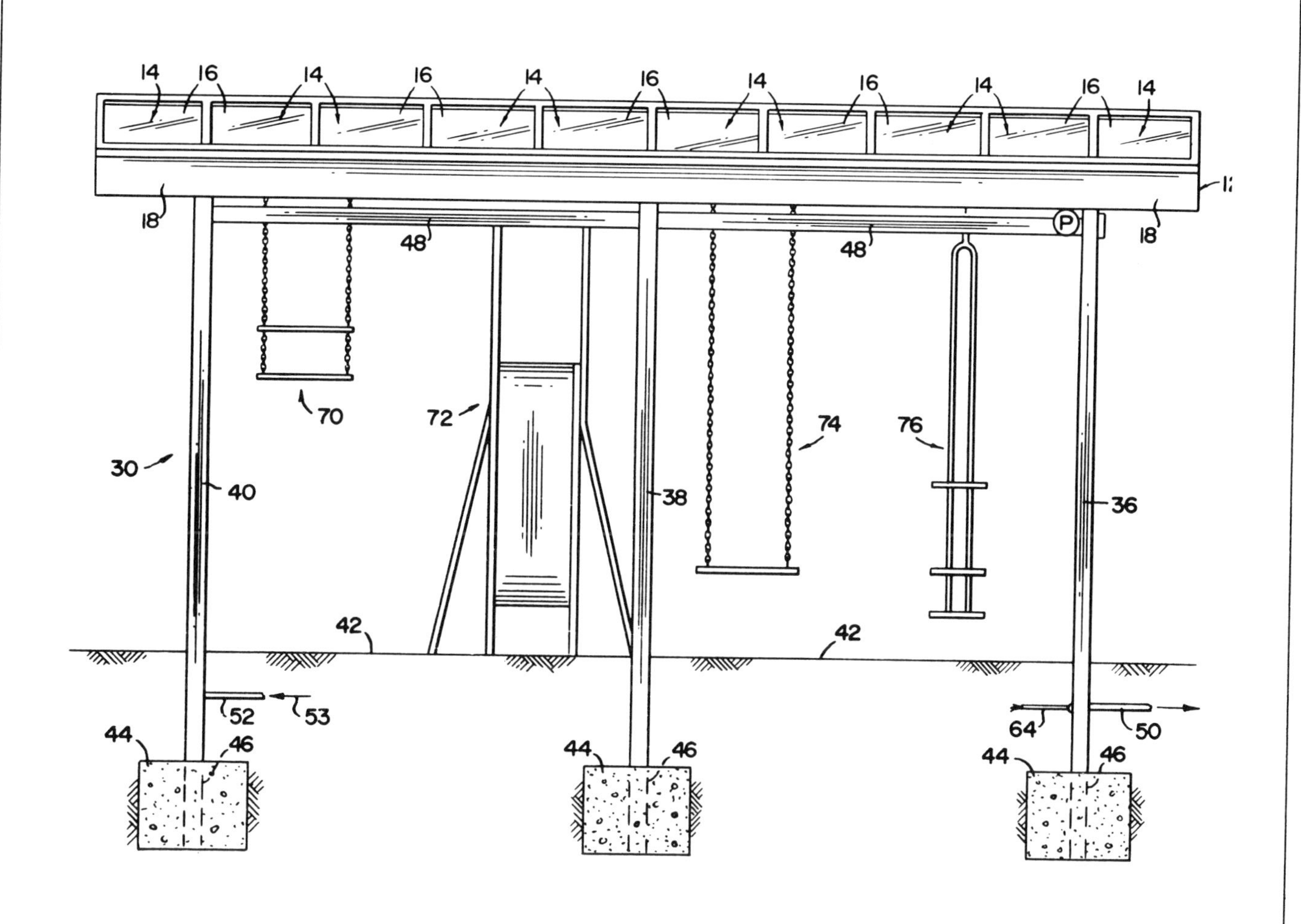

As shown by Netherton, homeowners can save space and build a good-looking solar collector by incorporating it into a children's swing set. Collector panels (14) and (16) are mounted on supports (40) and (36). Children's swings, etc., are attached to the lower parts of the collectors at structurally reinforced points. Fluid enters inlet (52) and flows up tubes (40) into the collectors, where, after heating, it is pumped down tube (36) to exit point (50) for storage or immediate use.

HOT-AIR-TYPE SOLAR HEAT-COLLECTING APPARATUS

U.S. PATENT 4,108,155

ISSUED TO HISAO KOIZUMI ET AL., 22 AUGUST 1978.

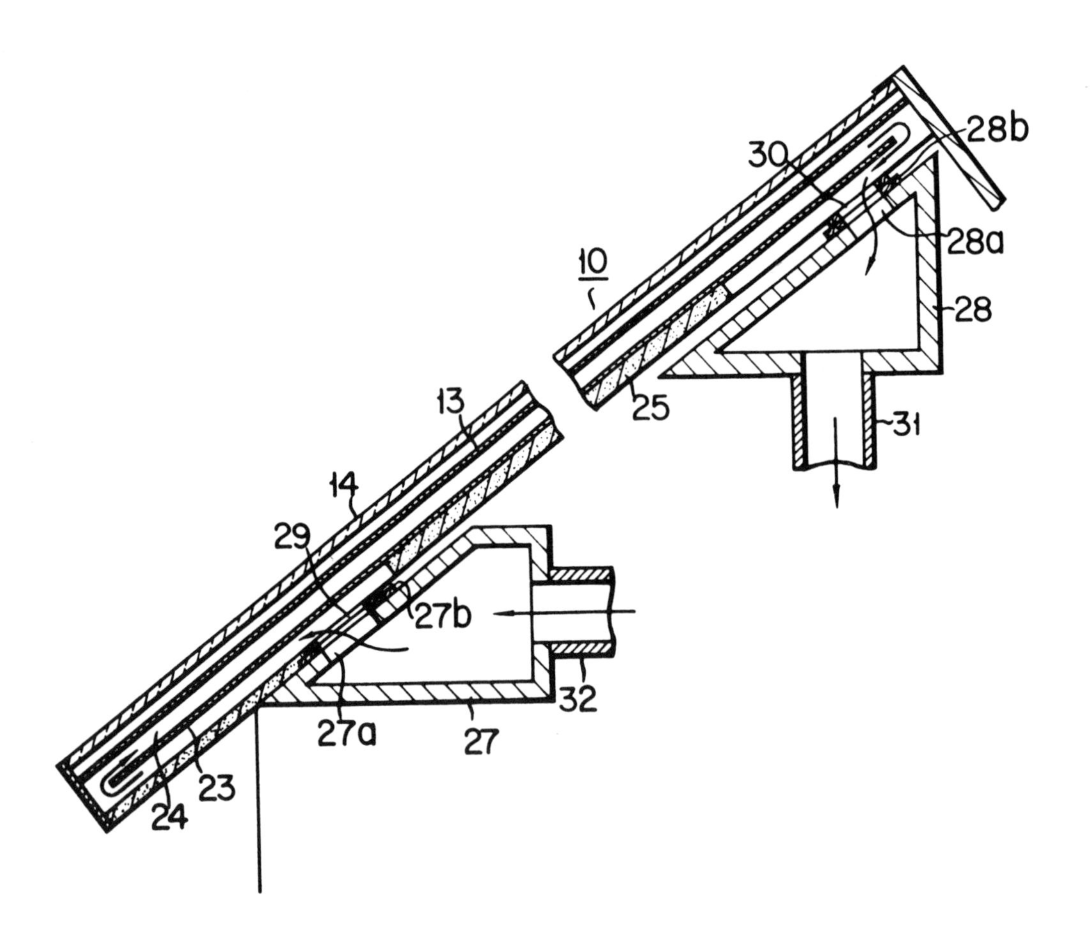

By combining a roof and solar collector into a single structure, the Koizumi system helps eliminate air- and rain-leakage problems present in many solar roof collectors. Air ducts (27) and (28) are angled to conform to the roof collector shape to provide better support. In operation, air to be heated flows over absorber plates (13), which are heated by solar energy passing through glass (14).

SOLAR-HEATED STOCK TANK

U.S. PATENT 4,108,156

ISSUED TO SPENCER B. SITTER, 22 AUGUST 1978.

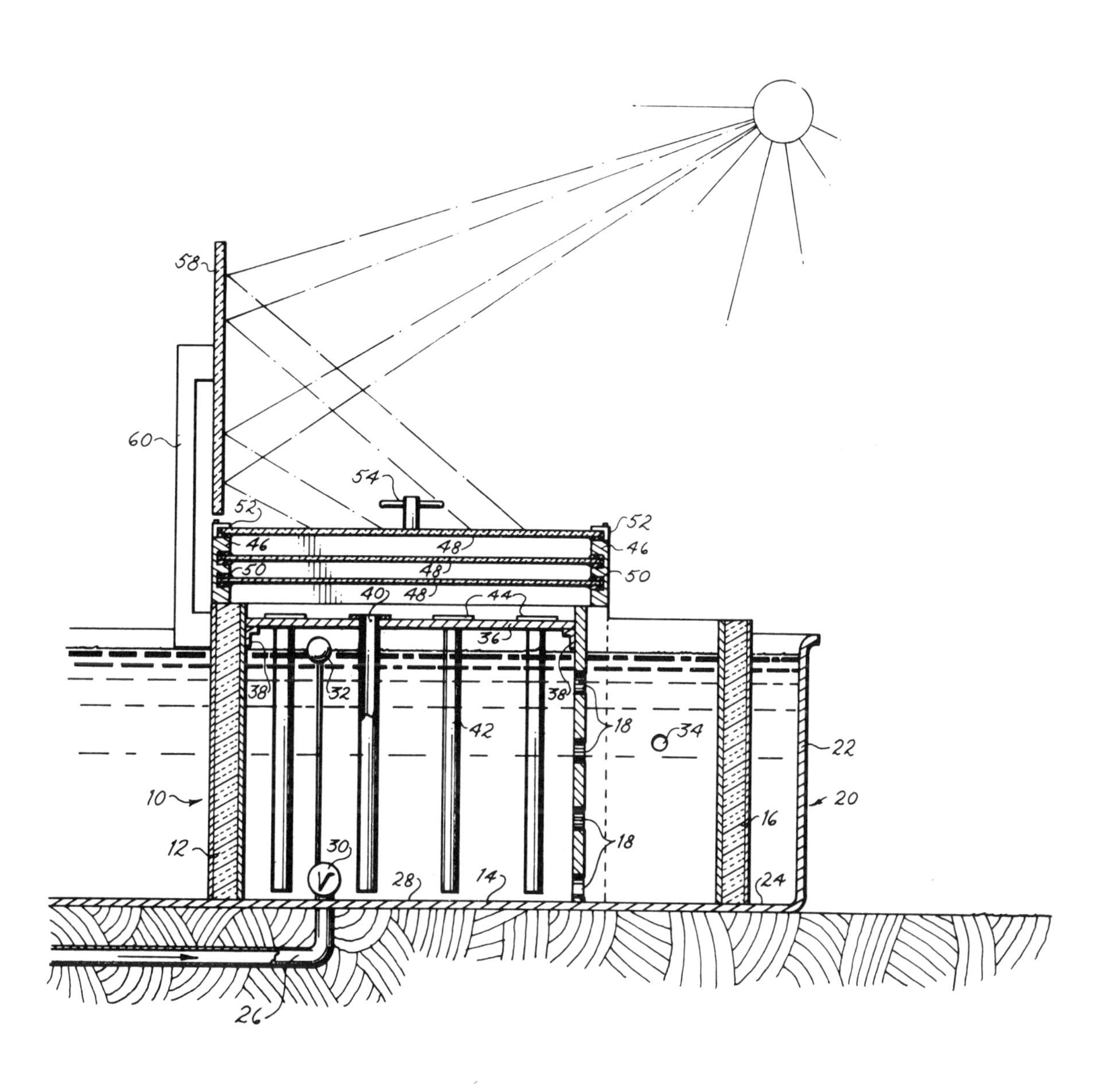

By using solar energy to prevent freeze-up of watering troughs for livestock, less wintertime care is required. Thus, cattle-watering troughs may be used at more remote or inaccessible locations. In operation, sunlight is reflected by plate (58) onto glass covers (48). Air under the glass covers (48) is heated and this heat is transferred down along tubes (42) to prevent ice formation or to melt any ice present. The sun-warmed trough water mixes with the rest of the water through holes (18).

SUN-TRACKING SOLAR-ENERGY COLLECTOR

U.S. PATENT 4,111,184

ISSUED TO JAMES C. FLETCHER, 5 SEPTEMBER 1978.

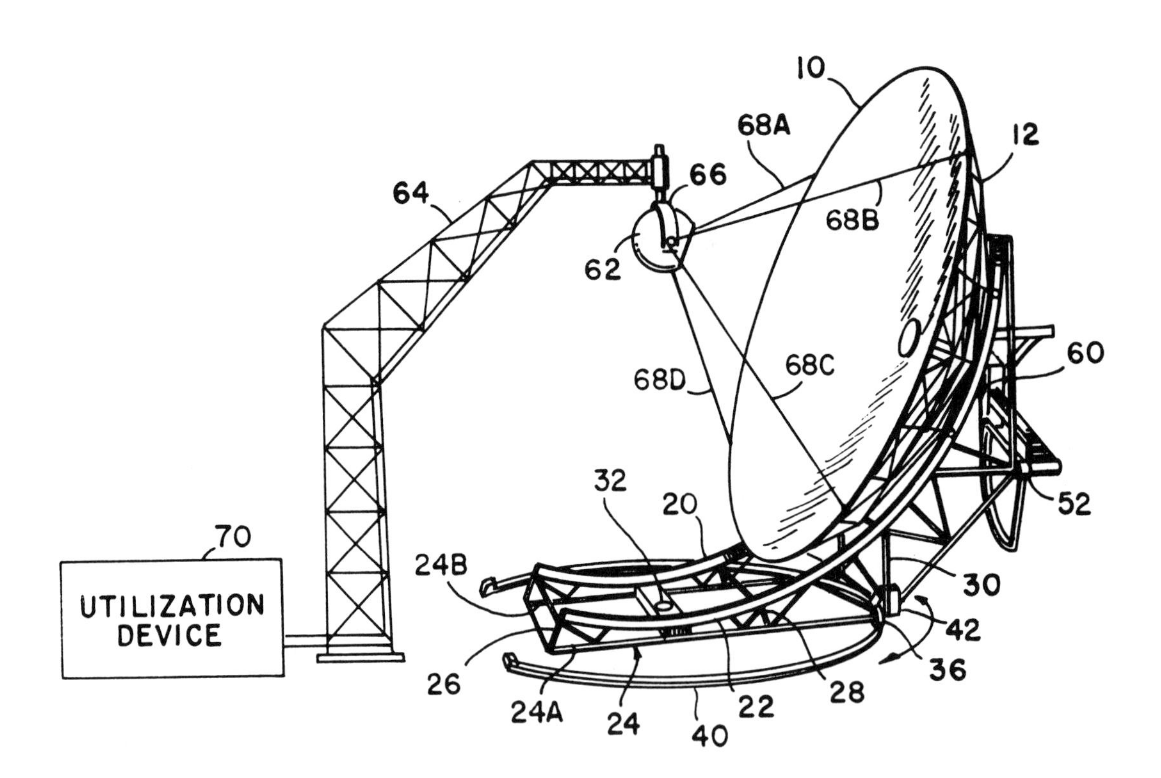

This sun-tracking collector system is designed to follow the sun by moving along metal tracks. In use, parabolic reflector (10) moves along track (20) and rotates about lower track (40). As reflector (10) is moved, it turns collector (62) about its own pivot point by means of connecting supports (68). Heat energy absorbed by the collector (62) is then piped to be used at utilization device (70).

SOLAR HEATING SYSTEM

U.S. PATENT 4,111,185

ISSUED TO FREDERICK R. SWANN, 5 SEPTEMBER 1978.

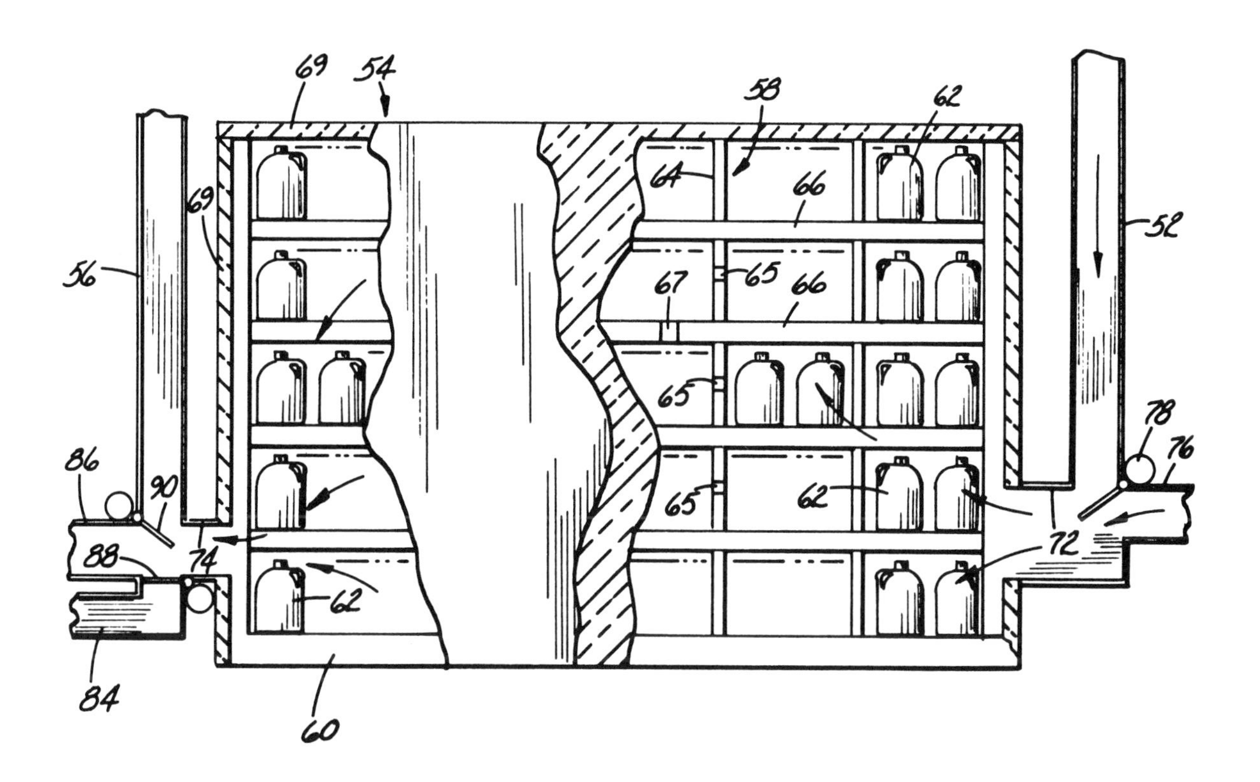

Ordinary plastic soda or milk containers can be used to store solar-heated fluids. In Swann's system, storage container (54) houses shelves (66). The plastic containers (62) sit on these shelves and are filled with fluid to absorb heat from solar-heated air passing through duct (52). Depending upon the setting of exit dampers (90) and (88), the air in storage chamber (54) may flow through duct (86) for home heating, to exhaust duct (84), or to return duct (56) to pick up additional solar heating energy. One major advantage of the Swann system is that a large amount of water can be stored in containers (62) without the cumbersome sealing and leakage problems present in a large tank that contains a single mass of water.

EXTRUDED METAL SOLAR-COLLECTOR ROOFING SHINGLE

U.S. PATENT 4,111,188

ISSUED TO JOHN A. MURPHY, JR., 5 SEPTEMBER 1978.

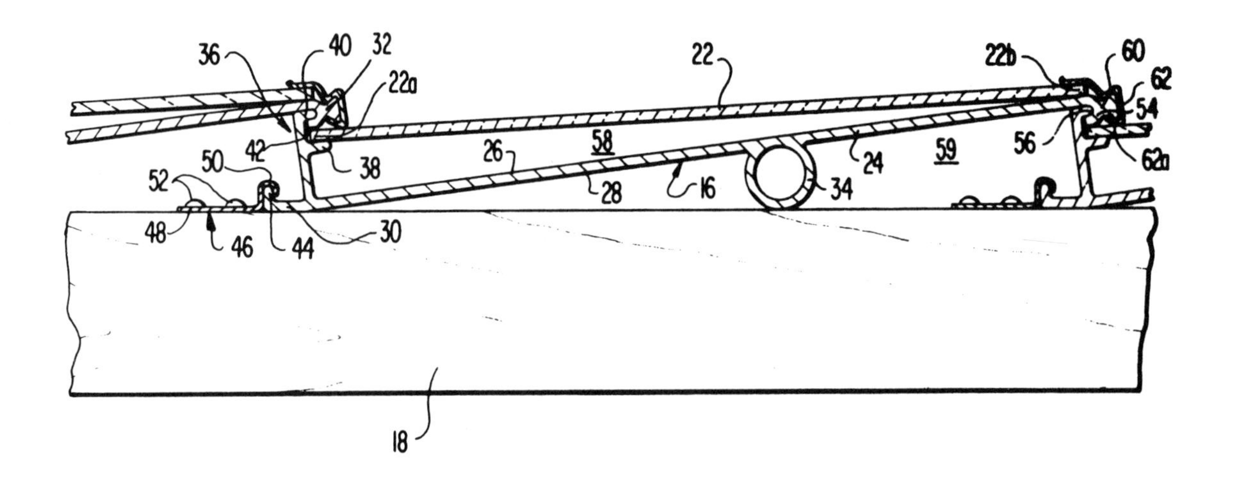

This solar collector has a pleasing appearance because it is more of an integral part of a building than the box-shaped collectors that merely rest upon a roof. In Murphy's design, roofing shingles are interlocked at end pieces (40) and (32). Clear upper surface (22) admits solar rays to chamber (58) and onto a black coated-aluminum layer (26). A fluid-carrying tube (34) is attached to the lower part of plate (26). Since aluminum is an excellent heat conductor, heat absorbed on plate (26) is passed readily to tube (34) and the fluid within it. Sealed air space (58) helps to insulate the system.

SOLAR-ENERGY COLLECTION DEVICE

U.S. PATENT 4,116,224

ISSUED TO RAYMOND R. LUPKAS, 26 SEPTEMBER 1978.

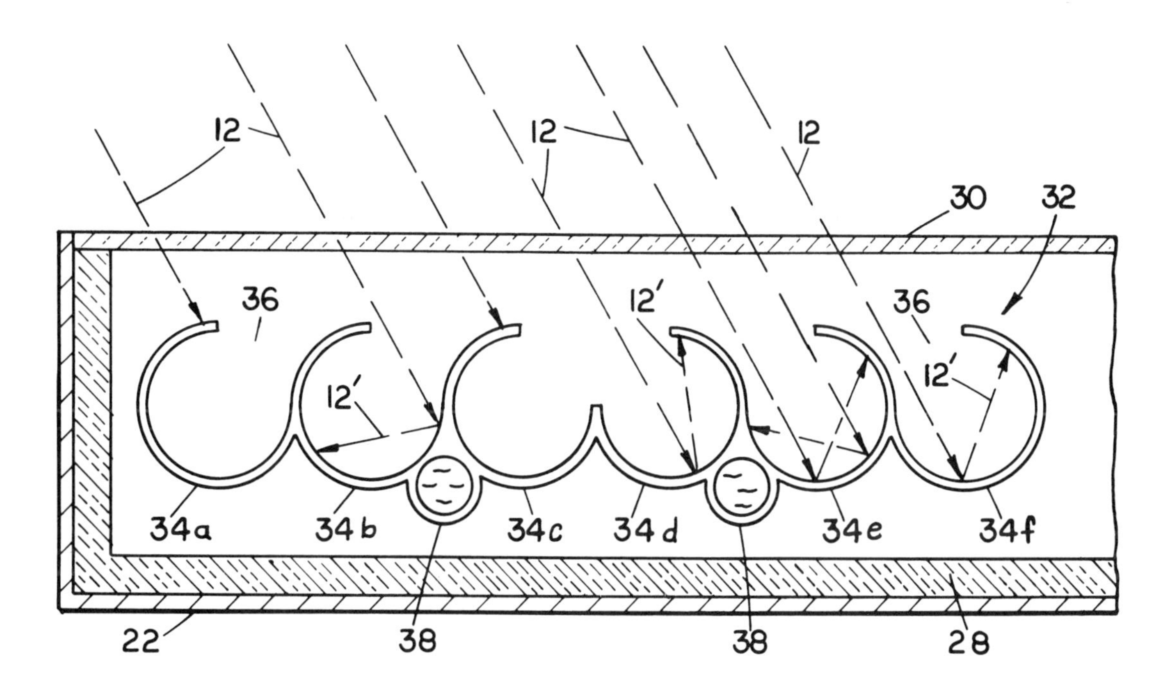

The Lupkas device illustrates the "tracking" of the sun throughout the day without the use of moving elements that require motor power. Collector plates are formed in a way to keep reflecting solar energy for more heat absorption by plates (34). The plates (34) are made of a high-heat-transferring metal to allow efficient heat transfer to fluid-carrying tubes (38). Base (22), insulator (28), and transparent cover (30) complete the system.

FURLIKE PHOTOTHERMAL CONVERSION SURFACE

U.S. PATENT 4,117,829

ISSUED TO DANIEL GROSS ET AL., 3 OCTOBER 1978.

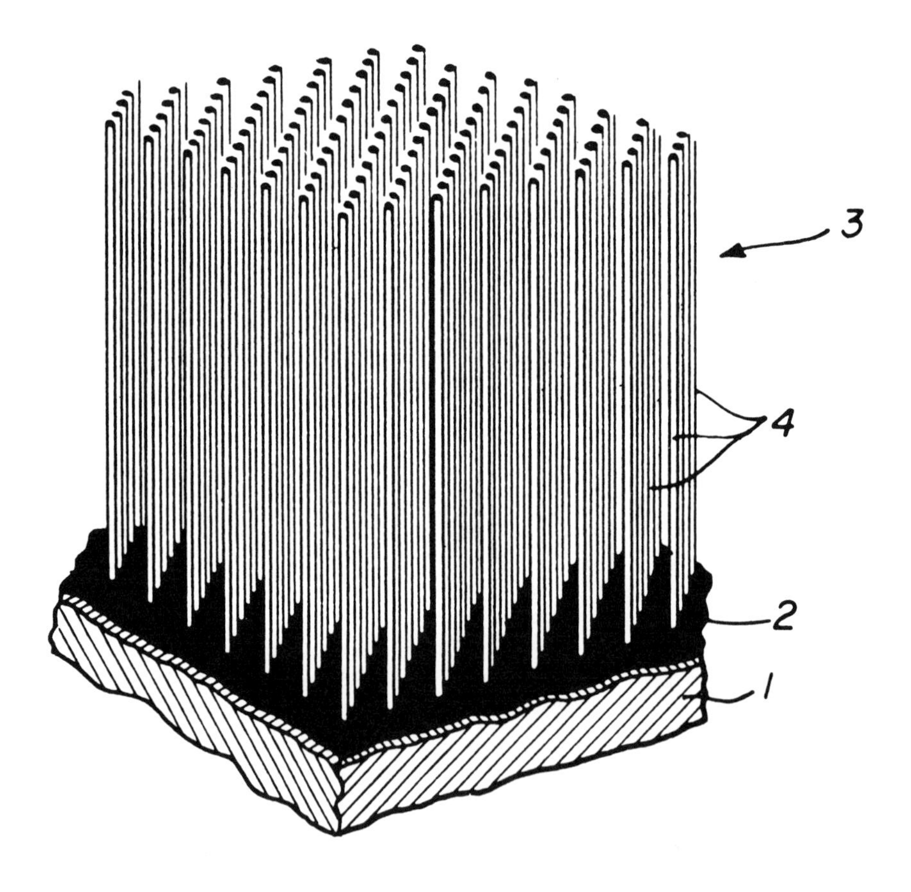

This design uses multiple glass or plastic fibers to absorb solar energy and transmit it to collector members (1) and (2). These glass or plastic fibers (4) serve much the same function as human hair, i.e., they protect the collector (2) from wind and other elements. Thus, heat losses from the collector (2) are reduced and the system performs more efficiently.

ENERGY COLLECTOR FOR COLLECTING SOLAR ENERGY AND THE LIKE

U.S. PATENT 4,117,831

ISSUED TO JAG M. BANSAL ET AL., 3 OCTOBER 1978.

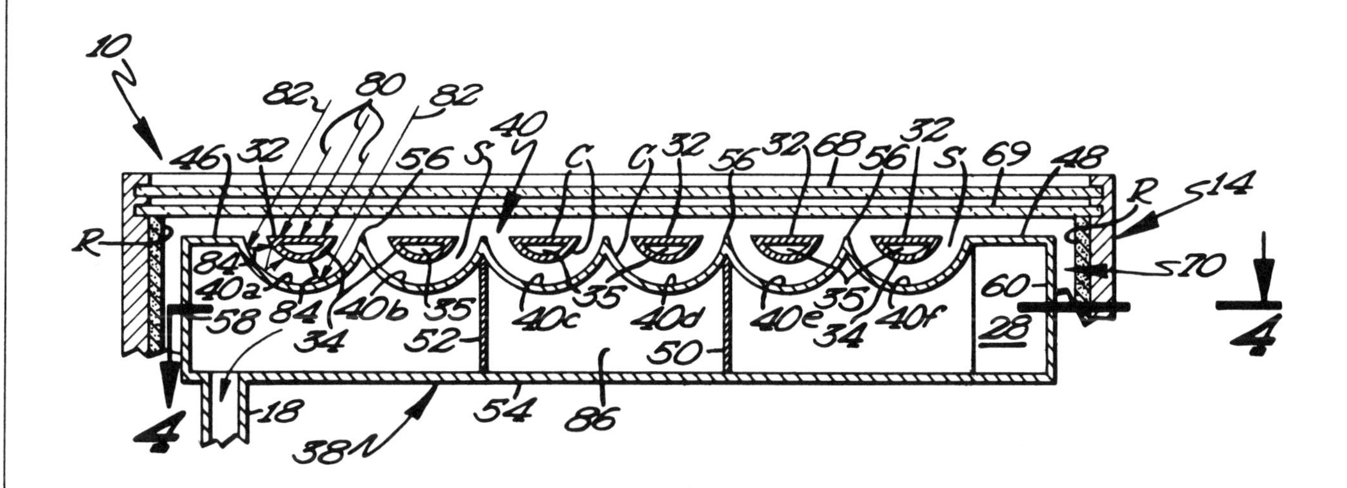

The shape of sun-receiving tubing can improve collector efficiency. As shown in the Bansal patent, two transparent insulating plates (68) and (69) seal sun-heated air within collector (10). Solar rays (80) strike the flat top surface of heat exchanger tubes (32). That part of light bypassing the flat tops (32) is redirected by reflectors (40) to a lower curved edge (34) of the tubes. Such an arrangement allows fluid circulating through the tubes at (35) to be more uniformly heated, since sunlight strikes the tubes from both top and bottom. Beneath the reflectors, in chamber (86), air or water may be circulated to pick up further heat energy.

BUILDING WITH PASSIVE SOLAR-ENERGY CONDITIONING

U.S. PATENT 4,119,084

ISSUED TO ROBERT E. ECKELS, 10 OCTOBER 1978.

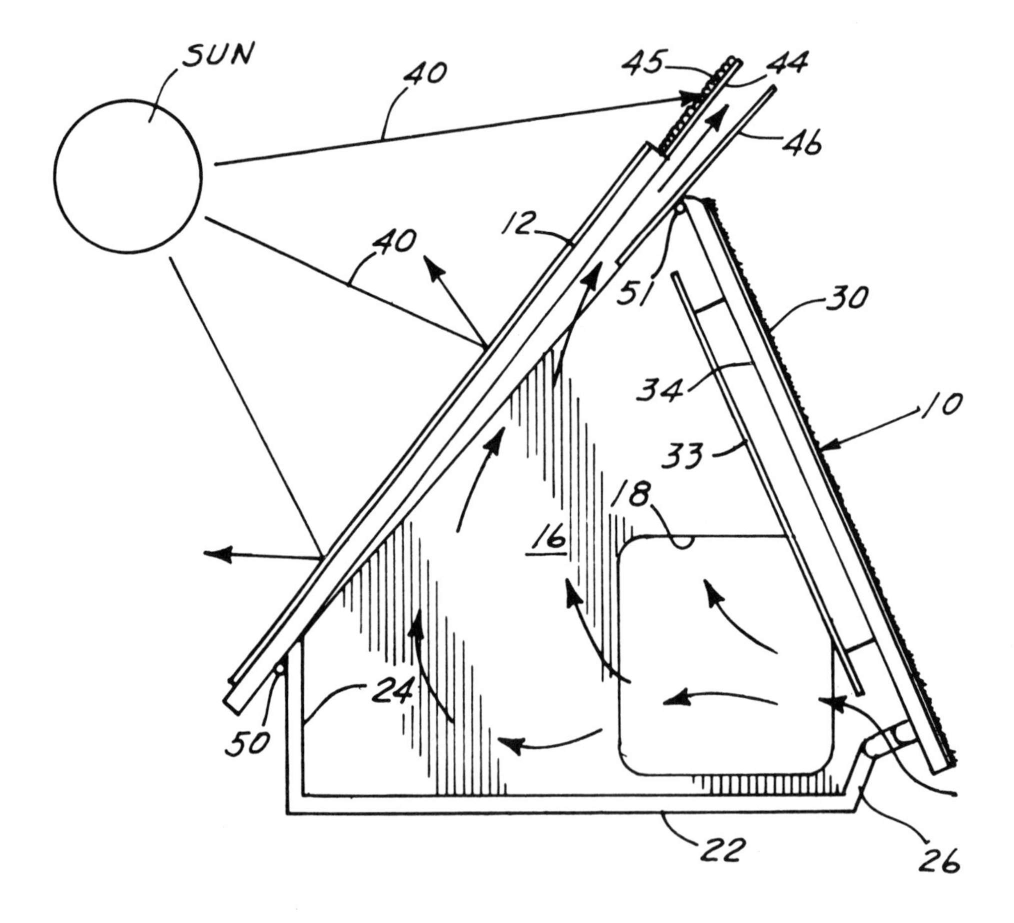

This patent shows the use of solar energy to both heat and cool a small building, such as a dog house. In summer months, solar absorbers (45) heat upper duct (46) and create an upward ventilating air flow via lower duct (26). Reflecting surface (12) directs unwanted solar energy away from most of the structure. In winter months, the small building is rotated so that solar heat-absorbing surface (10) faces the sun for an interior heating effect.

SOLAR-ENERGY COLLECTOR

U.S. PATENT 4,128,095

ISSUED TO JESS W. OREN, III, ET AL., 5 DECEMBER 1978.

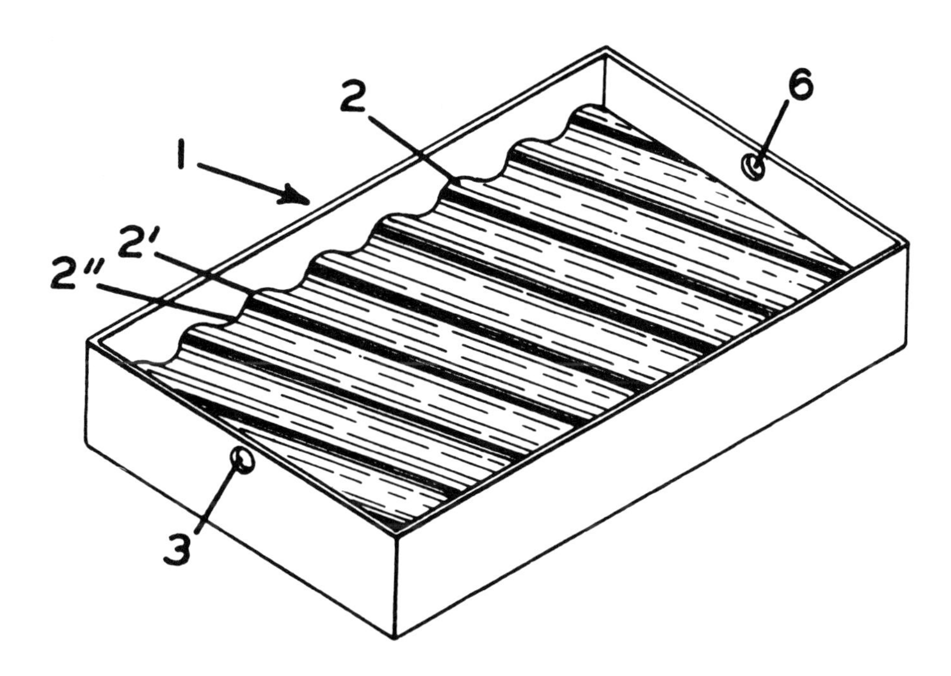

This invention varies a fluid flow pattern to increase the efficiency of a solar collector. A fluid flowing in a turbulent manner will absorb more solar heat energy than will a smooth-flowing fluid. The system uses an angled corrugated absorbing surface (2) to create a turbulence effect in the fluid that passes over the surface through ports (3) and (6). The fluid is heated both by solar energy passing through a glass cover and by heat that has been absorbed into surface (2).

SOLAR-ENERGY COLLECTOR

U.S. PATENT 4,129,117

ISSUED TO LAWRENCE HARVEY, 12 DECEMBER 1978.

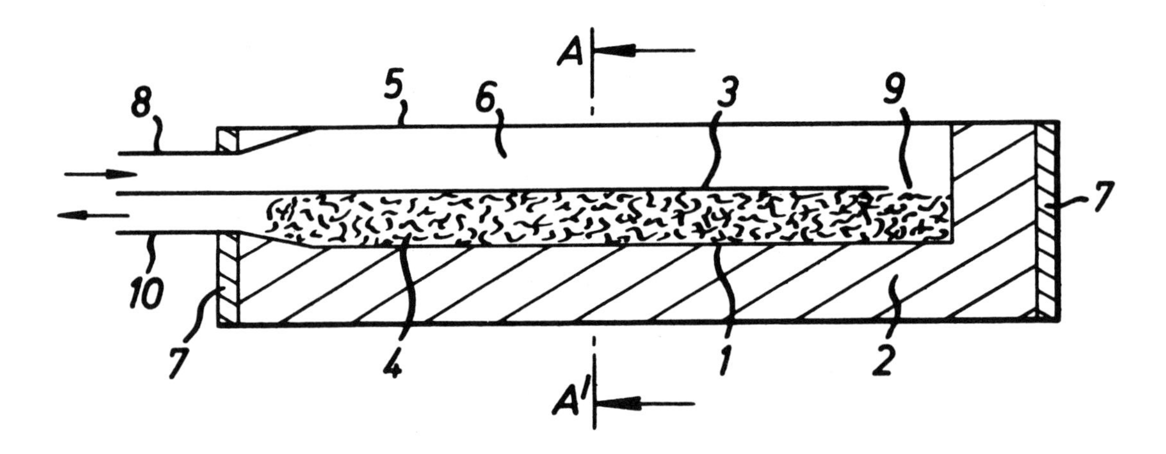

Harvey shows another invention that uses a fibrous absorbing material to enhance solar heat collection. In this system, air to be heated flows into duct (8) and under transparent member (5). It then flows down through passage (9) into duct (1), which contains loosely packed fibers that have been sun heated. The large contact surface area of the fibers and the turbulent flow path created allow the air to pick up more solar heat energy before leaving the collector at (10).

METHOD AND APPARATUS FOR PROVIDING DIFFERENTIAL TEMPERATURE THERMOSTAT FOR SOLAR HOT-WATER SYSTEMS

U.S. PATENT 4,129,118.

ISSUED TO KERRY O. BANKE, 12 DECEMBER 1978.

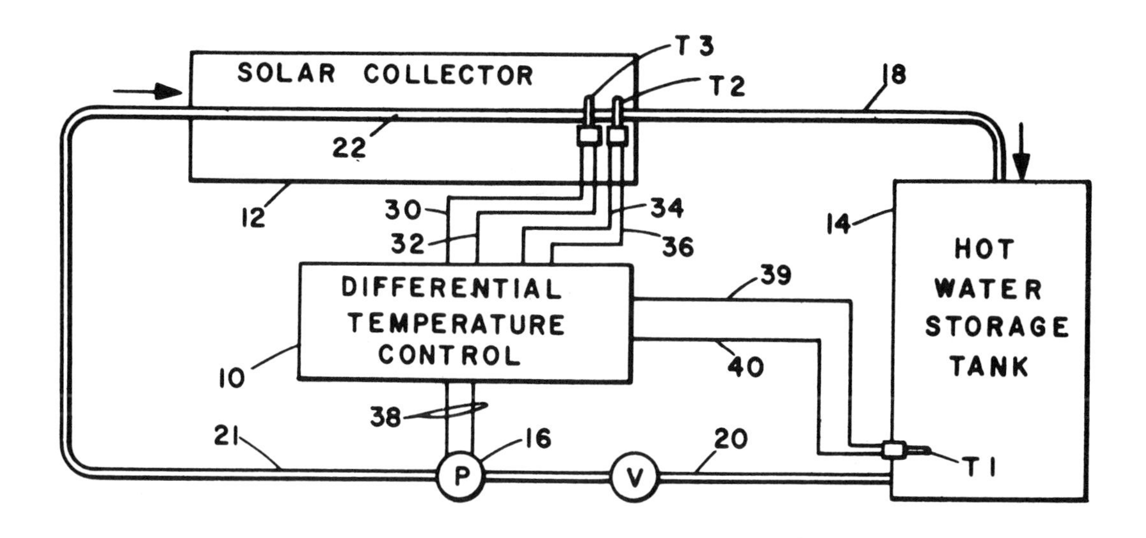

Automatic control is an important feature in many solar collector systems, as illustrated by Banke. In operation, temperature sensors at (T1) and (T2) generate signals that are fed into control circuit (10). When the heat at (T2) is sufficiently greater than at (T1), the pump (16) is turned on to permit flow of sun-warmed water from the collector (12) to the storage tank (14). Temperature sensor (T3) will not allow pump (16) to be activated if it senses a collector temperature below a certain value, thus preventing operation on cloudy days or in cold weather.

THERMAL-ENERGY TRANSFORMER

U.S. PATENT 4,135,367

ISSUED TO ROBERT A. FROSCH, 23 JANUARY 1979.

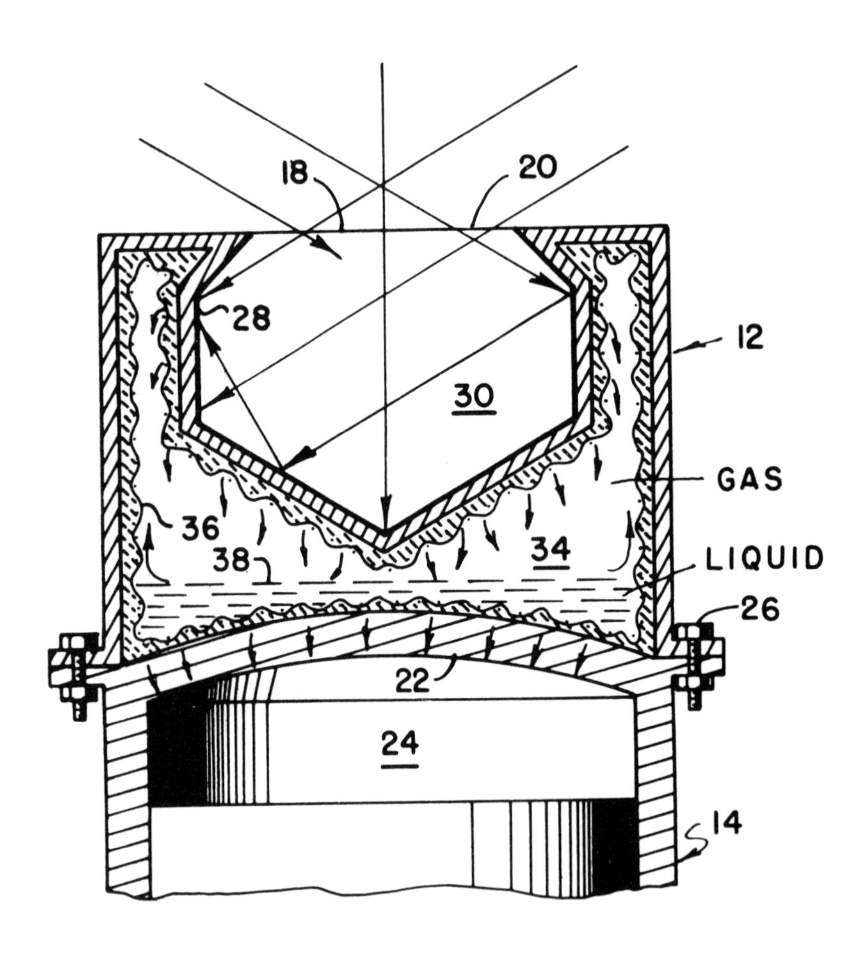

A heat transfer device known as a "heat pipe" is used in many solar-powered systems to transport sun energy. Frosch shows the use of a heat pipe to transfer solar energy to a heat engine. Solar energy entering chamber (30) heats and evaporates fluid on that part of a wick (36) that surrounds chamber (28). The evaporated fluid passes to lower liquid level (38) where it condenses and gives off heat through wall (22) to the heat engine chamber (24). Liquid is returned via wick-pumping action upward along member (36) to the evaporator section, and the process is repeated.

SOLAR HEAT STORAGE AND UTILITY SYSTEM

U.S. PATENT 4,136,668

ISSUED TO ARIEL R. DAVIS, 30 JANUARY 1979.

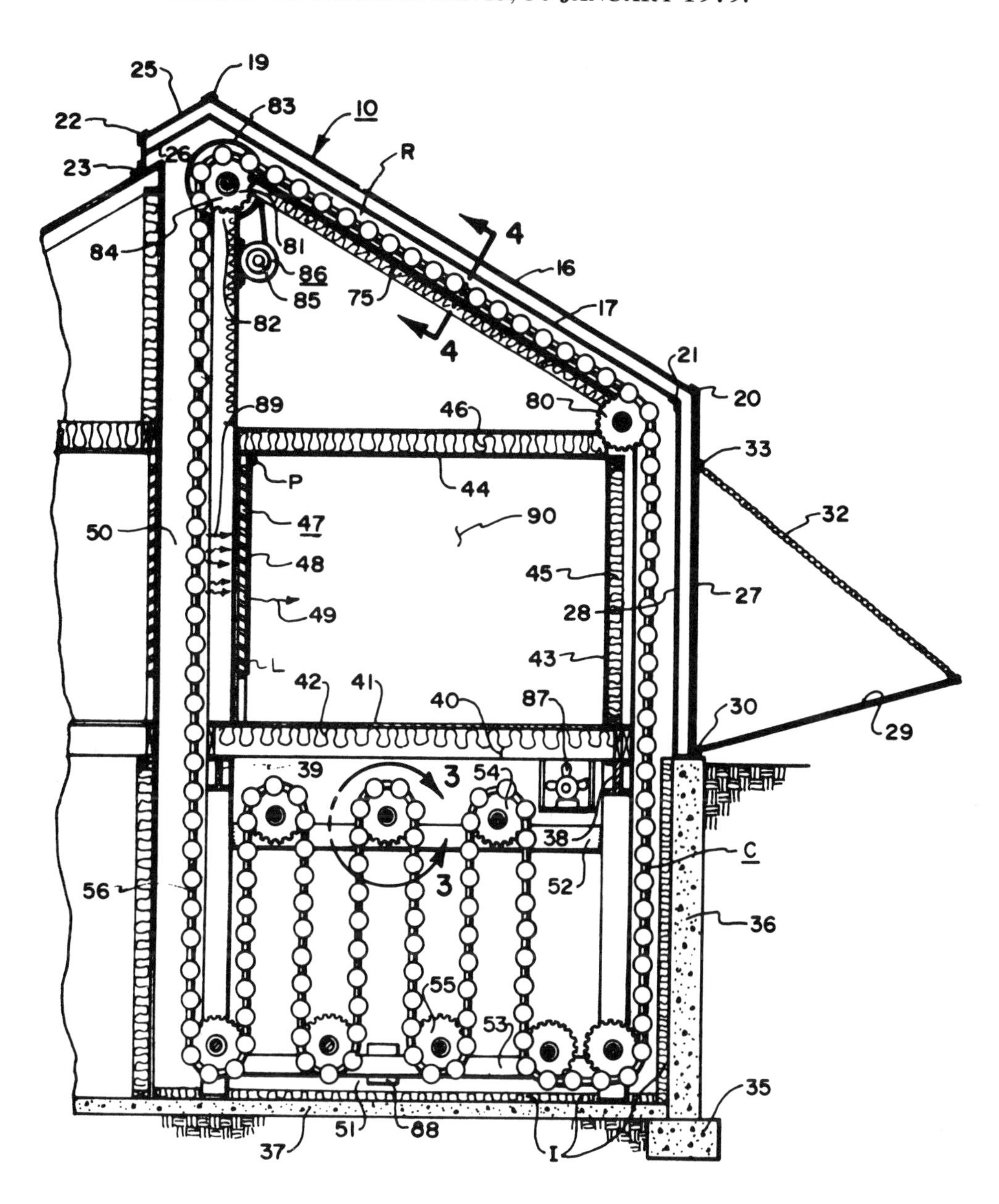

The Davis system demonstrates the use of movable solid containers, instead of flowing liquid or air, to accomplish a solar heat pick-up function. The solid containers are closed tubes that are mounted on sprocket chain (56). The tubes each contain a heat storage and transfer fluid and are moved by the chain beneath collectors (16) and (17) to pick up solar heat. As the chain moves the tubes down base member (75), they are rotated to more efficiently absorb solar heat energy. The heated tubes may give off heat directly to rooms (90) or may transfer heat to the lower storage tank (36) for use at a later time.

COMBINED HEAT PUMP SYSTEM AND ICE-MAKING SYSTEM

U.S. PATENT 4,142,678

ISSUED TO EDWARD W. BOTTUM, 6 MARCH 1979.

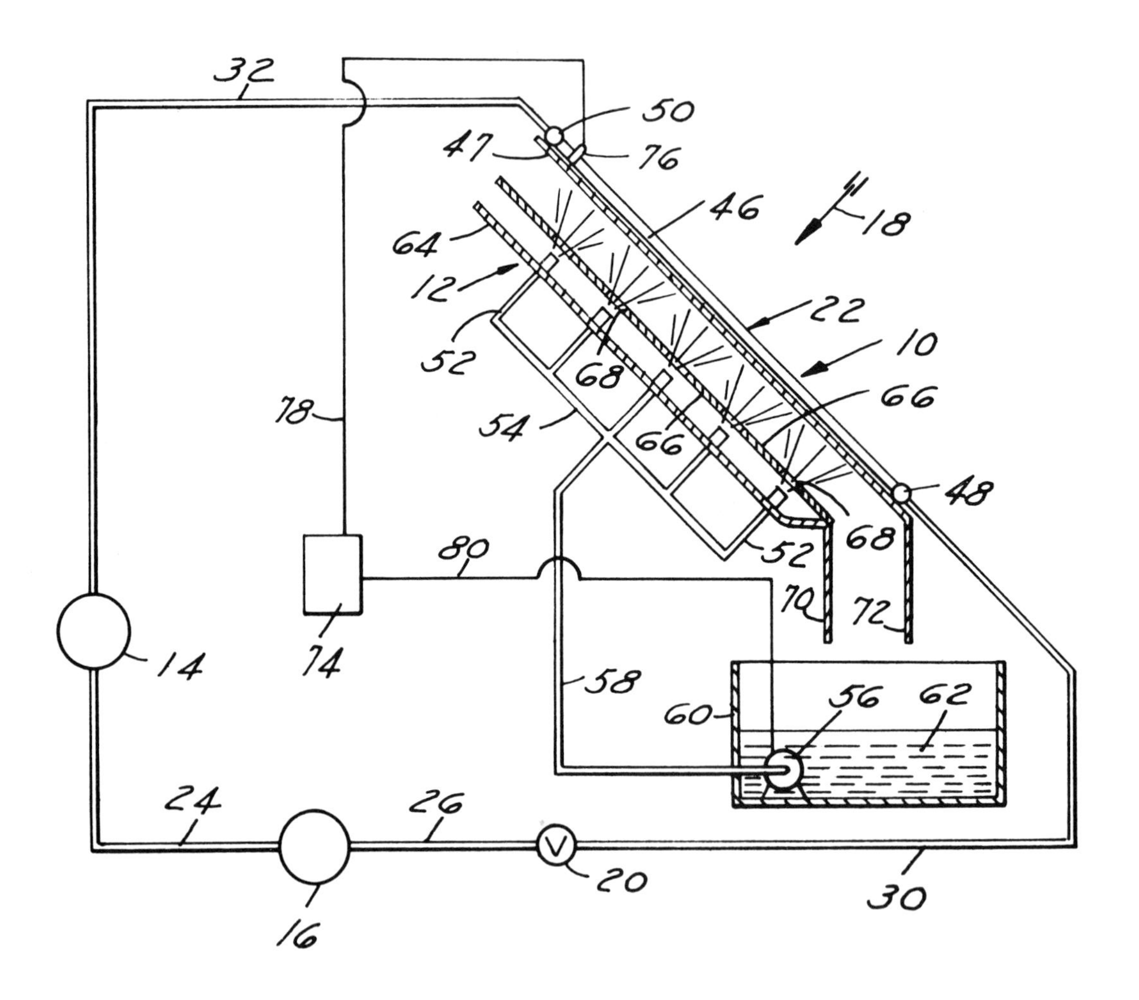

A solar collector panel can be used for more than receiving sunlight, as this patent to Bottum demonstrates. It shows the use of solar energy (18) in an evaporator (22) of a heat pump, used in many areas of the country to heat homes. The heat pump includes compressor (14) and condenser unit (16). Also included, for use at times when solar energy is unavailable, is a water spray system (54) that coats plate (22) with a film. The film is frozen and dropped into chamber (60) from which it may be removed for ice usage.

Index